New Complete Guide to Ferrets

NEW COMPLETE GUIDE TO FERRETS

JAMES MCKAY

Quiller

Copyright © 2012 James McKay

First published in the UK in 1995
by Swan Hill Press, an imprint of Quiller Publishing Ltd
as *Complete Guide to Ferrets*

This edition first published in the UK in 2012
by Quiller, an imprint of Quiller Publishing Ltd

British Library Cataloguing-in-Publication Data
A catalogue record for this book
is available from the British Library

ISBN 978 1 84689 131 1

Typeset by Phoenix Typesetting, Auldgirth, Dumfriesshire
Printed in China

Quiller

An imprint of Quiller Publishing Ltd
Wykey House, Wykey, Shrewsbury, SY4 1JA
Tel: 01939 261616 Fax: 01939 261606
E-mail: info@quillerbooks.com
Website: www.countrybooksdirect.com

DEDICATION

To Jane – still the best thing that ever happened to me – truly;
and to my son, James Thomas ('Tom') McKay, in the hope that his love of, and
interest in, ferrets and all animals continues throughout his life.

About the Author

At the age of 16, James McKay was introduced to the delights of ferrets and ferreting by a gamekeeper, and was immediately hooked. He has remained fascinated by both ferrets and the sport of ferreting, and has done much to aid the promotion of the animal and the sport. Annoyed by the misleading public image of ferrets, he determined to put the matter right, and has been giving talks and demonstrations throughout the UK, under the auspices of The Ferret Roadshow, since 1982. In 1992, he formed the UK's National Ferret School whose aims are aptly summed up by its Latin motto, *Ut docendo fortuna viverrarum melior fiat* (Improving the ferret's lot through education).

Internationally recognised as a leading authority on ferrets, zoologist James McKay lives in Derbyshire, with his wife Jane and their son, Tom, both of whom share his enthusiasm for ferrets. Previously chief executive of the UK's National Federation of Zoos, he is a Scientific Fellow of the Zoological Society of London, and an independent zoological and environmental consultant. He writes articles on ferreting, shooting, dog training, falconry, pet care, wild animals, zoos, green issues and other animal-related subjects for a wide range of publications, broadcasts on radio, and regularly appears on television, both as a presenter and an informed expert guest.

Contents

The author giving an arena display at a game fair – ferrets are always popular at such events. (Nick Ridley)

Introduction

Ferrets have for many years been extremely popular in such countries as Australia, New Zealand, the UK and USA, and also mainland Europe and Scandinavia, as working animals. However, people are also realising the potential of this intelligent, clean and easily trainable *mustelid* as a 'different' pet, as a 'fancy' animal, or for use in the fast-growing sport of ferret racing, as well as for the more traditional uses to which the animal has been put over the years.

In the USA, where ferrets cannot be used for hunting in many states, they have long been favoured as household pets, and command quite large sums when offered for sale. Regular shows and other ferret events are organised throughout North America on a regular basis by ferret clubs and societies.

The same situation exists in Australia (where ferrets help reduce the huge number of rabbits which plague the country), several European countries, New Zealand, and all of the Scandinavian countries. In addition, large numbers of ferrets are bred – particularly in Australia and New Zealand – for their pelts (fitch fur).

In the UK, the general public have not been slow to pick up on these trends, partly owing to the fact that many people will always want a pet that is different from the norm. Ferret shows and races are held throughout the country, mainly – but not exclusively – at country shows, where they have proved extremely popular and a huge crowd puller. Many ferret clubs have sprung up around the UK in recent years, and now have large memberships.

A similar state of affairs exists in Europe, with ferret clubs and societies existing in almost every country. All of these clubs organise ferret shows, races and other such events, and hunting (rabbits, rats, etc.) with ferrets is also extremely popular in those countries where it is legal.

Ferrets (and many of their relatives) have been with us as pets, fancy animals, and working companions and allies for thousands of years. People love them and people hate them. However, one thing that is common is that, regardless of their own opinions on the ferret, people are still fascinated by this remarkable animal.

The subject of countless music-hall and bad-taste jokes, the ferret is a much maligned and misunderstood animal. Far from being the nasty, spiteful, untrustworthy, dirty animal that some would have us believe, the ferret is clean, intelligent, and will never bite the hand of a caring owner.

As a pet, the ferret is just as friendly as any cat could ever be, as faithful as any dog, and more interesting than any rodent or cage bird. As a working animal, the ferret shines out, as this is what nature intended for the animal. Anyone – man, woman or child – who owns and works ferrets is guaranteed sport without equal, not to mention a regular supply of good, healthy meat from the rabbits which working ferrets will hunt with such vigour.

Within the pages of this book, I intend to put right many misapprehensions, untruths and nonsenses that are spoken (and written) regarding ferrets. The book is designed to ensure that everyone who has an interest in these delightful and under-rated animals is given all of the most up-to-date information available, and that every aspect of ferrets and ferreting is fully explained to all readers.

In the preparation of this book, I have had the assistance, help and guidance of many people, although naturally any mistakes are my own. I would like to express my thanks and appreciation to everyone who has helped in the writing of this book.

I would especially like to thank my son, Tom, and my wife, Jane. Tom shares my enthusiasm for ferrets and ferreting, although he saw far less of me than either of us wished, while I hammered out my manuscript on a keyboard. Jane has cheerfully endured many years of ferrets and ferreting, and the fact that I have been somewhat remiss with my attention to the household, especially the garden and the decorating.

Now the finished item is here, and I hope that all of those who have helped me so much over the years find the contents useful and, dare I say it, instructive. My heart-felt thanks go to all mentioned, and the many hundreds who are not, but have still supported and encouraged me in my labours.

James McKay
Derbyshire

Chapter 1

Origins

Where did ferrets originate?

You could be forgiven for believing that this is a simple question to answer; the truth is that no one can agree – not even the 'experts'. When, why and where were they domesticated? There are many theories put forward as to the origins of the ferret and each has its adherents. Although I can offer no conclusive proof either way, I have my own ideas, and it is always interesting to examine some of the recorded facts, along with a few of the many ideas which abound concerning the pedigree of the ferret.

The first historical mention of a ferret that I can find is in the Bible. In Chapter Eleven of Leviticus, the ferret is listed as one of the animals which Jews were forbidden to eat. However, in some versions of the Bible, the same word is translated as weasel or even lizard, and so this reference is obviously controversial.

Aristophanes mentions the ferret in his comedy *The Acharnians,* written in about 450 BC, and Aristotle lists the ferret in his *Historia Animalium,* written about 320 BC. However, these references too are considered to be somewhat controversial, as scholars disagree on the correct translation from the Greek.

The first universally accepted reference to the ferret that I can find is that by Strabo, a Greek historian and geographer. In his book *Geographica,* written in about AD20, Strabo writes of an animal in Libya which was said to have been bred in captivity for hunting rabbits. He states that the animal was either used to bolt the rabbits or, if this was not possible, then the animal would hang on to the rabbit while the 'ferret', complete with captured rabbit, was dragged out on the end of its lead. If this sounds difficult to believe, then you will have even more difficulty accepting the claim that, as this animal was fastened on a line and always muzzled, it held its quarry in its claws while being dragged out on the line.

I am inclined to believe that the animal did exist (although I doubt it was actually our ferret), but that Strabo, like many modern people, did not fully understand how the 'ferret' was actually used. Today, members of the Ruafa tribe, in Morocco, still hunt rabbits employing the services of a muzzled ferret-like animal.

Carolus Linnaeus, the Swedish naturalist and physician who, among other achievements, was responsible for 'inventing' a system of classification of all living things (first published in his botanical work *Systema Naturae,* in 1758), also seems to give credence to this reference by Strabo, as he listed the locality of the species he named *Mustela putorius* as Africa.

However, this raises several more problems in trying to establish the lineage of the ferret; neither of the animals from which most believe the ferret to be descended (the European polecat – *Mustela putorius* – and the Steppe polecat – *Mustela eversmanni)* come from Africa. However, as the skeletal features of these two species most closely resemble that of the ferret, very few people would argue that one or more of these animals is not the ferret's ancestor.

Some authorities, however, do choose to argue, and state that the ferret is always white with red eyes (Linnaeus described the eyes as 'rubicund'), and that the animal must be kept in a warm cage, or else it will surely die from exposure, and if they escape, the ferrets will be dead within forty-eight hours. The polecat, argue these people, is an entirely different animal, as it can withstand cold and is never white. These 'facts', argue our desk-bound 'scientists', 'prove beyond a doubt' that the ferret's ancestors originated from the southern hemisphere. I have even read a book written by 'the world's leading authority on ferrets' which claims that it is unnatural and wrong for a ferret to be fed meat. This author recommends that ferrets are fed a herbivorous diet. The same author also states that ferrets love to cuddle and sleep with rabbits.

Other 'facts' which these people give in their published works, include the ideas that ferrets can never be tamed, are always untrustworthy, and their behaviour is always erratic, making them totally unmanageable and completely unsuitable as pets.

Those of us with real, first-hand experience of ferrets, however, will know that these statements about the ferret's inability to withstand the cold are utter rubbish, as the ferret is quite capable of withstanding very harsh weather conditions (although it does need a dry and draught-free area into which it can retire – in common with all other land mammals). We also know that the ferret is bred in many colours and hues, from an exact replica of the European polecat, to pure white, and every shade in between.

We know too that ferrets can live in the wild (or go feral to use the proper terminology). Feral colonies of ferrets are to be found in almost every country throughout the world; in the UK the Island of Mull (in the Inner Hebrides) has a strong colony of feral ferrets, while in New Zealand, the ferrets have become a danger to the native fauna.

Ferrets were introduced to New Zealand in 1882, in an attempt to control the rabbit population in that country. However, the ferrets soon found that there was easier prey than rabbits. New Zealand, until that time, had no endemic small carnivores, and consequently, the indigenous fauna had developed a unique lifestyle; many birds had evolved as flightless and ground-nesting, and these were easy prey and tasty morsels for the ferrets. Today, over twenty endemic bird species in New Zealand are suffering from the unwanted attentions of the feral ferrets on the islands. These threatened species include the kakapo, and the national bird of New Zealand, the kiwi.

The ferret can, of course, be tamed and is *not* the sneaky character of legend. I find it vexing that, in almost every novel (or even movie) about animals, ferrets or their relatives are always portrayed as being the bad guys. Such novels include *The Wind in the Willows,* and the movie *Who Framed Roger Rabbit?;* and how many times have you read or heard the description of a nasty, sneaky thief who would sell his own mother for the price of a glass of Scotch, as being 'ferret-like'?

Perhaps this would be a good place to mention the names given to ferrets and their offspring; the males are known as hobs, although many erroneously call them dogs, while in Norfolk, England, they are known as jacks. The females are called jills, although again some wrongly refer to them as bitches, and the young (under three months) are called kits; this latter term comes from the offspring of the poleCAT – poleKITTENS. A castrated male is known as a hobble, and a vasectomised male as a hoblet. The collective noun for a group of ferrets is a business; some authorities seem to have made up their own name, and often the term 'mess' is used. This is totally incorrect. I have researched the subject and, in a book written by Dame Juliana Bernes, *The Boke of St Albans,* which contains a list of 'the companys of beestys and fowleys' *(sic,* the term given is a *'besynes of ferrettis';* there can be no mistaking the updating of that term to 'a business of ferrets'. I am also certain that terms such as this are given for good reason; they often accurately describe the nature of the species. Anyone who has ever witnessed two or three ferrets turned loose together will tell you that they certainly get on with their business of ferreting things out, without any prompting or encouragement.

Of course, the very name ferret will help establish the animal's perceived character, as it comes from the Latin *furonem,* which means thief (hence the word *furo,* often put as the third part of the ferret's scientific name). Today, ferret is also used as a verb, meaning to search diligently, to remove from a hiding place, or to draw out by shrewd questioning; a military vehicle, used for reconnaissance and scouting purposes, also has the name 'Ferret'. Built by Alvis in the UK, the company obviously chose the name very carefully, and the vehicle, until quite recently in service with the armies and security services of many countries world-wide, lives up to the reputation of the real ferret.

Other names for the ferret and its wild relatives include fitch, fitchet, poley, stinkmart, stinkmarten, foulmart, foulmarten, foumart and fulimart. Most of these names help signify the smell of the ferret; even the most ardent fan of the animal cannot but agree that the ferret does have a very characteristic smell.

'Poley', 'polecat ferret', 'fitch' and 'fitchet' were terms originally given to a hybrid of the wild polecat and its domesticated cousin, the ferret. Today, however, the terms are used to describe any ferret which resembles the polecat, i.e. has polecat markings. Legends about hybrids of ferrets and polecats abound; some will tell you that the two animals cannot produce offspring. Others will tell you that any offspring from such a mating will be infertile; some will tell you that kits from polecat to ferret matings are easy to spot, but impossible to tame. Although all of the foregoing 'facts' are totally wrong, almost everyone will tell you that they have gained this knowledge from first-hand experience; I have never yet met an 'experienced ferreter' who does not claim to have had (or still has) a *true* polecat. When questioned, however, the source of this 'polecat' is always obscure and unbelievable. I am convinced that, in ninety-nine point nine per cent of cases, the 'true polecat' is simply a large, dark-coloured ferret with polecat colouring. I refer to all ferrets of any coloration kept in captivity simply as ferrets.

I believe that the ferret is simply a domesticated form of the European polecat, *Mustela putorius.*

Having said that, I too claim to have true polecats. I obtained my first specimen, a hob named Rasputin, from a British zoo; its ancestors had been taken into captivity around the turn of the last century, having been bought from a gamekeeper who had trapped them. Since that day, no other animals have been introduced to the colony, and so they are as true a polecat as we are likely to find. It is worthy of note that, when scientists carried out genetic investigations on wild polecats, they found that many of them were hybridised with feral ferrets.

The name polecat gives an indication of the ferret's propensity for stealing; it comes from the French *poulet chat,* meaning 'chicken (killing) cat'.

No one with first-hand experience of ferrets could possibly deny that ferrets *do* make excellent pets, provided that the owner takes the time and makes the effort to handle their charges regularly, and treat them well. My own ferrets will come to my call, walk along at my heels, and love having their tummy tickled. I allowed my son (when just three years of age) to handle and play with all of our ferrets when we were around to supervise, and he was never bitten or attacked in any way by the ferrets – even when he interrupted or joined in their games of 'chase'. The same ferrets, however, will willingly and effectively flush rabbits from their underground home.

Pliny the Elder (Gaius Plinius Secundus), the Roman scientist and historian, wrote many books, some of which survived his fate (he was killed in the eruption of Mount Vesuvius in AD79). The subjects covered by these books were geography, astronomy, and natural history. One of these books, *Natural History,* mentions both the ferret and the rabbit, and there are many more such references in his book.

Isidore of Seville, a writer and missionary, who was appointed as Bishop of Seville, Spain, in AD600, wrote many important books, including his *Ethymologiae,* the model for later medieval encyclopaedias. In his book *Patrologie,* written about AD600, he wrote of the ferret's use in rabbit hunting.

It is not known for certain when the first 'ferret' (or more correctly semi-tame polecat), was introduced into Great Britain, but I'm sure that it must have been linked with the introduction of another foreign invader – the rabbit. The rabbit, whose history is intertwined and inexorably linked with that of the ferret, was originally indigenous to Spain and North Africa but, by the end of the first century AD, was firmly ensconced in southern France.

Until the Roman period, the rabbit was restricted to an area encompassing the Iberian Peninsula, North Africa and southern France; some scientists believe that, if the rabbit had not been introduced to Europe, then it would have shared the same fate as the European pika, and become extinct. So although most species can blame man for their untimely demise and suffering, the rabbit is one of a select group of mammals (which include the horse and the dog) whose widespread distribution and success owes much to the activities of *Homo sapiens.*

The Romans were responsible for the rabbit's spread from Spain, but they did not make any attempt at domesticating the species; instead they chose simply to catch the rabbits and fatten them for the table, or leave them to live in the wild until needed for their culinary delights. For convenience sake, the Romans kept many rabbits in artificial areas (i.e. warrens), until they were taken for the pot. However, as everyone knows, rabbits are accomplished burrowers, and it was not long before many made their escape, thus colonising new areas. Most authorities (although not all) agree that the rabbit did not reach Britain's shores until Norman times.

The selective breeding of rabbits was probably **first** carried out by monks, who regarded the newly-born and unborn rabbit kittens as 'not meat'; this convenient classification enabled the monks to eat the rabbit kittens during 'fasts'. However, the monks' rabbits, kept in walled and paved areas for convenient harvesting, also escaped, and this helped to spread the wild population even more.

To try to combat these escapes, many land owners in the Middle Ages established rabbit colonies on islands, where the animals (and their spread) could easily be controlled. Ships also carried rabbits with them on long voyages, and left small colonies on the islands that they stopped at, as a food source to tap into every time they were in the area. This latter practice led to devastation on some islands, with the indigenous species being wiped out owing to the habits of the rabbit. One of the best (or worst) examples of this was Round Island, near Mauritius.

In the 1970s, the island was in a diabolical state, owing mainly to the rabbits and goats put there by man, presumably for food supplies. Many of the endemic species were in danger of imminent extinction and, had it not been for the sterling efforts of Gerald Durrell and the Jersey Wildlife Preservation Trust, they would have joined the dodo and countless other species driven down the one-way road to oblivion that is extinction – and extinction is for ever. By 1994, Round Island was, thanks to Gerald Durrell and his helpers, on its way to recovery, following extensive trapping and shooting of the foreign invaders, thus clearing the way for the island's indigenous species to re-colonise the land. This has only been possible because of captive breeding in establishments like Jersey Zoo; if ever you were in any doubt about the need for zoos in the twentieth and twenty-first centuries, this surely is proof.

We will probably never really know whether it was the Normans or the Romans who were responsible for the introduction of the rabbit to the United Kingdom, but the introduction was undoubtedly for their use as a source of food. I have no doubt, however, that whoever was responsible for the introduction of the rabbit to Great Britain, was also responsible for the first use of the domesticated polecat or ferret.

By the Middle Ages, the ferret was widely used for hunting the rabbit, and many references to this practice still exist in historical records. Queen Mary's Psalter of 1340 clearly shows very well-dressed ladies ferreting, although the work probably stylises the true scene. In his *Livre de Chasse* (*circa* 1387), Gaston, Comte de Foix, shows drawings of ferreting taking place, complete with a muzzled ferret and several pursenets. This idea of muzzling ferrets in order to make them unable to kill is one which persists even today, and yet the practice renders the ferret extremely vulnerable to attacks by rats, which often live in rabbit burrows. Although one can almost forgive 'the ancients' for their ignorance, I find it impossible to forgive modern-day ferreters for this sin. Gaston's work also states that rabbits were only hunted by members of the fur trade.

In 1221, at Termez, a place on the banks of the River Oxus, about 160 miles south of Samarkand (Afghanistan), legend has it that the infamous Genghis Khan used ferrets for hunting rabbits, while in a court roll of 1223, we find the first concrete evidence of ferrets and their use against rabbits in Britain, and there is documentary evidence that a ferreter was attached to the English Royal Court in 1281.

In the thirteenth century, rabbits, and so obviously ferrets, were important to the Church and its establishments, as borne out by the many references to ferrets which belonged to quite high-ranking churchmen of the time. Rabbits and ferrets were also of great importance to the landed gentry, who did not wish for the peasants to share this bounty. In an attempt to keep ferrets out of the hands of would-be poachers, a

A fifteenth-century Franco-Burgundian tapestry, showing an idealised ferreting scene. Note the purse-nets and dogs. (The Burrell Collection, Glasgow Museum and Art Galleries – The Ferreters' Tapestry)

law was passed in 1390, which limited the ownership of ferrets to those with a minimum salary of forty shillings per year, the equivalent of a very highly paid executive today. This law indicates that poaching was becoming a major problem, and was obviously designed to exclude ferrets from the lower classes.

The first mention of the white ferret is from 1551, in Conrad Gesner's *Historiae Animalum;* this was a book dealing with all quadrupeds, and published in Zurich. The ferret's colour is described by Gesner as 'the colour of wool stained with urine', an excellent description. In 1421, John Lydgate, a Benedictine monk at Bury St Edmunds, England, later prior of Hatfield Broadoak, English poet and friend of Geoffrey Chaucer, wrote the poem *The Seige of Thebes.* One line in this poem mentions the animal's red eyes, another obvious reference to the white (albino) ferret.

The lords of the manor in Medieval England thought so highly of the rabbit that they kept specific fields for them to live in, and built warrens to try to ensure that the animals stayed in that area. The men in charge of these artificial warrens were responsible for the protection of the warren and for ensuring that the lord had a regular supply of rabbit meat for his table. These men were referred to as 'warreners', and it is from this that the surname 'Warrener', 'Waren', and 'Warrender' came.

The household books of Lord William Howard of Naworth Castle (1618–1633), make interesting reading, and give us a wonderful insight into life for a warrener. The warrener's house had a tiled roof and a brick floor. He was supplied with a 'wallet' (a carrying bag) for his ferrets, seven yards of coarse cloth for his bed, eight yards of cloth for his blankets, a hank of yarn for mending his nets, a paddle staff with iron on it (an implement for digging out laid-up ferrets, etc.), various traps – for rabbits and their predators – and a ferret line. In 1622, a ferret room was built for his ferrets (the ferrets were kept in barrels and later hutches in this room, and all of the warrener's ferreting equipment was also stored in the room). Add all this up, along with his not insubstantial wages, and it can be seen that the warrener was a servant of great importance to his master.

Although most people now accept that the rabbit was introduced to Britain by the Normans, some still maintain that it was the Romans. These people argue that the rabbit was used to supplement the soldiers' rations and to ensure that, wherever Rome's Legions marched, there would be food available. The Romans were renowned for taking food animals with them wherever they travelled. One favourite of their senior officers and nobles was the edible dormouse *(Glis glis);* the animals were kept

Queen Elizabeth I with her 'pet ferret'. The author believes that much of the 'ermine' used to trim the robes of royalty was, in fact, ferret fur – fitch. (The Ermine Portrait of Queen Elizabeth I, by kind permission of the Marquess of Salisbury)

in earthenware jars until they were fat enough for the pot (although I doubt that even then they would provide more than a couple of mouthfuls of meat), and cooked with fruit and nuts before being served. According to Pliny, this practice was banned in 14 BC, because the Roman senators believed that the eating of such delicacies showed too large a degree of ostentatious luxury, and was an affront to the lower classes of Rome; at the same time shellfish and birds were also banned from the tables of the rich citizens of Rome.

The popularity of the ferret in the UK is graphically illustrated by the fact that at the turn of the nineteenth century, the British patent office was inundated with applications for registration of patents for ferret cages – a sure indication that the ferret was big business, and expected to get bigger. Their main use at this time was for rodent extermination – a practice that was then also very popular in the United States. According to US sources, tens of thousands of ferrets were raised and sold for this purpose every year at the turn of the twentieth century. The term 'ferretmeister' was used in America, and the US Department of Agriculture issued regular bulletins detailing – and encouraging – the ferret's use in rodent abatement.

Throughout its time with man, we have found many uses for the humble ferret. It has been used for hunting; in laboratories for influenza and cold research; fur (the ferret's pelt is known as 'fitch'); as a pet, as a fancy animal (i.e. for exhibition purposes), for ferret racing, and even in industry. The television cables at the Prince of Wales' first wedding were put into position by using a ferret; the animal was equipped with a harness, onto which was attached a light but very strong line. The ferret was then sent through the small pipes, etc. and, when it emerged at the other end, the line was attached to the TV cables, which were then hauled through. This practice is still used in industry for cables, telephone wires, etc.

Legends abound about other uses for the ferret; some claim that it is/was used in the United States space programme, while others claim that the customs and excise authorities in several countries are training the animal to detect drugs, and that the armed forces, police and security forces are training the ferret to detect explosives.

'Camel hair' brushes are made from ferret hair and *not* that of the camel, which in any case is woolly. A *putois* is a French brush made from ferret hair and used for painting glaze on ceramics.

The Ferret in the United States of America

The ferret was probably taken to North America (now the USA) by the first English settlers in the sixteenth and seventeenth centuries, and the animal has since had mixed fortunes in the 'New World'.

In the United States, in some states it is permissible to hunt rats with ferrets, provided one has the relevant licence. On the whole, US ferrets are kept as pets, and cost about twenty-five times as much in the US than they do in the UK. Consequently, veterinary surgeons in the US see ferrets far more often than do their British counterparts, although the American Veterinary Medical Association Council on Public Health and Regulatory Veterinary Medicine formally discourages the ownership of pet ferrets. As can be imagined, this has led to much controversy in the veterinary profession, all hinging on whether the ferret is a wild or a domesticated animal. Those against the keeping of pet ferrets argue that it is a wild animal, while those in favour argue that, since the ferret has been domesticated for over 2,000 years, it cannot possibly be considered to be a wild animal.

After the First World War, the ferret was used extensively in rodent control throughout the United States, and the US Department of Agriculture published regular bulletins on all aspects of this work. The widespread availability of effective rodenticides led to the demise of the ferret in professional rodent clearing operations, both in the US and in Great Britain.

Even before this time, ferrets were used for rodent extermination, and an excellent example of this is the Massachusetts Colonial Navy. This was America's Revolutionary War naval militia, and was formed on 29 December 1775, and reactivated in 1967, by an act of the Commonwealth of Massachusetts state legislature, to carry on the traditions of the unit. Ferrets were used on the eighteenth century ships to keep the rat and mouse populations on board under control, and this was recognised at a ceremony at Bristol Community College (USA) on 14 September 1986, when the ferret was officially proclaimed as the mascot of the Colonial Navy of Massachusetts. The first ferret with the title had the pet name 'Pokey', and it was possible to purchase cuddly Pokey toys.

Today, the ferret is looked on very differently in the US, and American ferret owners spend large amounts of money on ferret requisites – food, toys, shampoos etc., and there are companies that solely produce and/or sell goods for ferrets and their keepers.

However, not all US citizens are happy with ferrets as pets, and after a couple of nasty accidents involving ferrets 'attacking' children, several states and organisations are campaigning to have the animals banned, or labelled as a 'wild animal', which would also effectively ban the general public from keeping ferrets as pets.

Historically, the ferret has been categorised as a wild animal in every state except Alaska and West Virginia, and this has limited the habit of keeping these animals as pets. In some states, such as New York, special permits are required, while in California, the only ferrets that can be kept as pets are hobbles (neutered males). Georgia, Massachusetts, New Hampshire and South Carolina have already banned ferrets as pets, citing one reason as the lack of a suitable rabies vaccine for ferrets. The United States Humane Society lists the ferret as an 'exotic animal', and as such actively and publicly discourages the keeping of them as pets.

Because of attacks on children (there appear to be very few of these, but the ferret's opponents use them to their own advantage), some areas, including Carson City, Nevada, have banned the sale of ferrets to any household containing a child of three years of age or under. Despite the fact that under US legislation the ferret is now categorised as a domestic animal, Alaska, a state which had always allowed the keeping of pet ferrets, tried to ban them; a pet owner successfully appealed against this decision. Such appeals have also succeeded in defeating motions in other areas, including Maine and West Virginia.

I am sure that the ferret will be with us for many years, but ferret keepers through-out the world should carefully study the history of the ferret in the US, and in particular the moves taken (and still being taken) by individuals, voluntary organ-isations, professional organisations and even local and national government bodies to ban the keeping of pet ferrets.

THE FERRET AND THE FUR TRADE

As stated earlier, ferrets have been used to hunt rabbits for hundreds of years, and these rabbits would have supplied both meat and skins. The skins were made into many items of clothing and even into bags and footwear.

In the mid-nineteenth century, many people in mainland Europe kept and bred ferrets for their fur, which was called 'fitch'. The practice caught on in the USA in 1900, and there has been considerable interest in fitch fur production ever since in that country. Until the late 1980s, the popularity of fitch fur production was spread-ing rapidly, with fitch farms in Australia, Canada, the USA, New Zealand and many European countries. However, fashions change, and fitch, along with many other furs, fell out of favour in many countries in the late 1980s, although it would seem that the pendulum has again swung, and fur is once more being seen as acceptable, with some even making the claim that it is a 'green' (i.e. environmentally friendly) product.

It was (and still is) common practice to hybridise ferrets with European and Steppe polecats, to get better quality and nicer looking fur.

FERRET LEGGING

This so-called 'sport' involves the placing of ferrets inside one's trousers, and until recently competitions were held to find the 'world champion ferret legger', i.e. the one who could keep a ferret down their trousers for the longest period of time. Under the rules of the 'championships', all ferrets used must have a full set of teeth, and the human involved could not wear any underwear. Poor ferrets.

The origins of this horrendous activity were probably born out of necessity. As a poacher, one would not wish to advertise the fact that one had a ferret about one's person, and so carrying the animal in a bag or a box would be out of the question.

Instead, the poacher would have hidden the ferret about his person, possibly down his trouser legs. As it was common practice until very recently to stitch together the ferret's lips, in order to prevent it biting a human or killing a rabbit, the poacher would probably have been in no danger from the animal's teeth, although its claws could still have caused injuries.

Another theory is that, after working ferrets in the winter cold, the animal would have been in need of warming, and by placing it inside one's own clothing, this would have been achieved. I prefer to place my ferret in a weather-proof carrying box, along with a good supply of warm hay.

BIOMEDICAL RESEARCH

In the 1900s, the ferret's potential as a model for biomedical research was recognised, and it has been – and still is – used for research into a cure for human influenza, a malady that can also badly affect ferrets naturally. Today, the ferret is used as a laboratory (research) animal for physiology, virology, immunology, pharmacology, toxicology, and teratology.

SCIENTIFIC CLASSIFICATION OF THE FERRET

The ferret belongs to the weasel tribe, or *Mustelidae* ('seeker of mice', 'those who carry off mice', the name refers to the musty smell), and is thus related to such animals as the otter, badger and the skunk. It is also a carnivore, and thus related to cats (including lions and tigers), dogs (including wolves and foxes) and civets (meerkats and mongooses).

The *Mustelidae* is an extremely diverse family of mammals which includes terrestrial, aquatic and arboreal species. While most are true carnivores, some have evolved very specialised feeding. There are twenty-six genera, sixty-seven species and over 400 subspecies of mustelids recognised by scientists, and no doubt more will be added. They are endemic throughout the world, with the exception of Antarctica, Australia, Madagascar, New Guinea, New Zealand, most of the Philippines, Sulawsi, the West Indies, and most oceanic islands.

Almost without exception, mustelids are renowned for their fur, and many species have suffered from the fur trade, while many are of great economic significance in some communities; the best illustration of this is the sable *(Martes zibellina)*.

There are five subfamilies, viz: *Mustelinae* (weasels), *Mellivorinae* (honey badger or ratel), *Melinae* (badgers), *Mephitinae* (skunks), and the *Lutrinae* (otters). The pelage is either uniformly coloured, striped or spotted, with some species turning white in winter in the northern parts of their range (controlled by photoperiod). In most species, the body is long and slender; however, the wolverine and badgers are stocky. All have short ears, which are either rounded or pointed, and have short limbs with five digits on each. The claws are non-retractable, and are curved and compressed, with the badgers' claws being larger and heavier, as an adaptation for digging. In otters, the digits are usually webbed.

Males of all species in this family have a baculum (a bone in the penis), and most species have well developed anal scent glands. Skulls are sturdy, with short facial regions, and the dental formula is;

(i 3/2 – 3, c 1/1, pm 2 – 4/2 – 4, m 1/1 – 2) x 2 = 28 – 38

Enhydra is the sole genus with two lower incisors.

All mustelids are either nocturnal or diurnal, sheltering in trees, burrows or crevices, while badgers usually dig elaborate burrows, known as setts or cettes. The smaller, slender forms in this family move with a characteristic wave-like motion and, when excited, will bounce with stiff legs. The larger members of the family move with a rolling bear-like shuffle.

Mustelids are essentially carnivorous (otters eat mainly aquatic animals), although some species occasionally feed on plant material, while a few are omnivorous. All have highly developed hearing and sight, but tend to hunt by scent, and many species, including *Mustella, Poecilogale, Gulo,* and certain badgers, are storers of food. Many mustelids have anal glands capable of exuding a foetid secretion when threatened or attacked.

Delayed implantation occurs in many species of this family; although gestation is about thirty-five to sixty-five days, in *Lutra canadensis,* pregnancy can last as long as twelve months.

There is usually a single litter each year, although some species will produce a second litter if mated early in the summer. Offspring are self-sufficient at about eight weeks, sexually mature between one to two years, and can live from five to twenty-five years.

CLASSIFICATION OF THE MUSTELIDAE

ORDER....................*Carnivora* – 7 families, 93 genera, 231 species.

This order includes cats (including lions and tigers), dogs (including foxes, jackals and wolves), bears (including polar and grizzly), raccoons (including coatis and giant pandas), weasels (including stoats and polecats), civets (including meerkats and mongooses) and hyenas.

FAMILY....................*Mustelidae* – 26 genera, 67 species.

The *Mustelidae* are to be found all over the world, in every continent except Australia and Antarctica. Their habitat ranges from arctic tundra to tropical rainforest on land, in trees, in rivers and in the ocean.

Animals in this tribe range in size from the diminutive least weasel *(Mustella nivalis rixosa),* which measures fifteen centimetres head-to-tail with a tail length of three to four centimetres and weighs as little as thirty grams and is the smallest carnivore known to man, to the massive bulk of the grison *(Galictis vittata),* a mammal from Central and South America, which resembles a wolverine and can attain sizes of up to 120 centimetres head-to-tail with a tail of up to sixty-five centimetres and weighs up to thirty kilograms. One of the most recently described sub-species of the *Mustelidae,* is the Hainan small-toothed ferret-badger *(Melogale moschata hainanensis).*

SUB-FAMILY....................*Mustelinae* – 10 genera, 34 species.

These include the grisons, martens, mink, polecats and weasels. They are terrestrial hunters of small vertebrates, although some (particularly the martens) are excellent climbers.

Members of the Sub-Family **Mustelinae**

M. africana	Tropical weasel
M. altaica	Mountain weasel
M. ermininea	Common stoat (ermine)
M. eversmanni	Steppe polecat
M. felipei	Columbian weasel
M. frenata	Long-tailed weasel
M. kathiah	Yellow-bellied weasel
M. lutreola	European mink
M. lutreolina	Indonesian mountain weasel
M. nigripes	Black-footed ferret
M. nivalis	European common weasel
M. nivalis rixosa	Least weasel
M. nudipes	Malaysian weasel
M. putorius	European polecat
M. putorius furo	Domesticated ferret
M. sibirica	Siberian weasel
M. strigidorsa	Back-striped weasel
M. vison	American mink
Vormela peregusna	Marbled polecat
Martes americana	American marten
M. flavigula	Yellow-throated marten
M. foina	Beech marten
M. martes	Pine marten
M. melampus	Japanese marten
M. pennant	Fisher
M. zibellina	Sable
Eira barbata	Tayra
Galictis cuja	Little grison
G. vittata	Greater grison
Lyncodon patagonicus	Patagonian weasel
Ictonyx striatus	Zorilla, or striped polecat
Poecilictis libyca	Saharan striped weasel
Poecilogale albinucha	White-naped weasel
Gulo	Wolverine

THE BLACK-FOOTED FERRET

The black-footed ferret was once extremely common on the Great Plains in the United States, with the estimated population in 1920 being 500,000. However, throughout the twentieth century, it suffered as a result of the extermination – mainly by humans, but also by disease – of its sole prey, the prairie dog (*Cynomys spp*). The animals gradually became a rarity, and excitement spread among zoologists with the discovery of a small population in South Dakota, in 1964. Some of these ferrets were captured and taken into captivity, but no breeding took place. In 1974, all traces of this population vanished, and it was assumed by many that the black-footed ferret had passed into the oblivion that is extinction.

Then, on 26 September 1981, dogs on a ranch near Meeteetse, Wyoming, killed a ferret, and the rancher's wife was so impressed by the cadaver that the rancher took it to a taxidermist for mounting. Luckily, the taxidermist correctly identified the animal, and US Fish and Wildlife Service agents were alerted to the presence of what was to turn out to be a very large (possibly the largest ever observed) colony of black-footed ferrets. The agents succeed in trapping a ferret, which was fitted with a radio transmitter collar, and then released, and tracked it until it led the agents to the colony. After intensive field studies of the area by scientists, the population was estimated at 129 animals in 1984.

The excitement and elation of the discovery was short-lived, however, as the number of ferrets rapidly declined in 1985, as the result of a plague in their prey species, the white-tailed prairie dog *(Cynomys leucurs)*. As feverish efforts were made to control the plague in the prairie dogs, canine distemper was accidentally introduced into the population of black-footed ferrets (probably by the dog of one of the

Black-footed ferret. (Photos courtesy of the US Fish & Wildlife Service)

23

field workers), and this had a disastrous effect; by 1986, only 18 ferrets remained – the last of their species.

Scientists made a controversial decision, and captured all of the animals, placing them in captivity. Never before had such a drastic step been taken to help prevent an animal becoming extinct, but the need was great. Scientists worked hard, and finally managed to get the animals to breed, the first litter being fathered by a male known as 'Scarface'. Scarface went on to father many more litters before his death on 15 June 1992, and this fact caused many problems, as scientists tried to ensure that inbreeding depression did not occur.

By August 1992, there were over 240 black-footed ferrets in captivity, and a release programme was initiated, releasing selected animals back into the wild. Two areas were chosen, both in Wyoming; the Shirley Basin/Medicine Bow area, and the Meeteetse area. Scientists appealed to humans to avoid shooting prairie dogs and not to disturb the black-footed ferrets. They also asked that any dogs taken to the area should be fully vaccinated against distemper.

The ferret release sites are very small (about 1,500 acres), and it is to be hoped that humans can be persuaded to continue giving this magnificent animal the opportunity to recolonise the area – an opportunity it so richly deserves.

The black-footed ferret is the most endangered mammal in the USA, and is the only member of the weasel tribe currently listed as 'Endangered' by the International Union for the Conservation of Nature (IUCN); it is similarly listed by the USDI, and is on Appendix 1 of the Convention on International Trade in Endangered Species.

MUSTELID CHARACTERISTICS

The reproductive habits and characteristics of mustelids deserve some mention here. For most of the time, the sexes keep themselves to themselves and, when they do meet, they are often extremely hostile towards one another. It is only when the female is in oestrus (on heat) that a (temporary) truce is called. Then the male takes hold of the female by the scruff of the neck, often almost brutally, and proceeds to drag her around until he finds a suitable 'honeymoon site', where he mounts her.

Copulation is prolonged and repeated (it can last for over two hours) and this is facilitated by the baculum of the male (a bone in the penis). The mating often seems to be almost a fight between the male and female. This is probably nature's way of ensuring that only the strongest pass on their characteristics to the next generation since, if the male cannot 'overpower' the female, he will not be able to mate her.

In some mustelids, delayed implantation occurs. This is where the fertilised egg simply floats around in the uterus until being implanted in the uterus wall at a later date. This date can be anything from a few days to almost one year. (This is *not* the case with the ferret).

THE ATTRACTION OF THE FERRET

Ferrets are intelligent and clean; even though they do have a characteristic smell, they have excellent habits, always using the same area for defecating and urinating. They keep their fur meticulously clean, spending long hours grooming themselves. They are also entertaining to watch as they go through their everyday antics. Watching a ferret trying to break an egg will keep anyone enthralled for hours; I often wonder, at the risk of sounding anthropomorphic, if the ferret really *is* trying to break the egg, or simply playing with it first, before deciding to eat it.

Since 1982, I have spent much of the summer travelling around the UK to various events with my 'Ferret Roadshow', putting on displays with my ferrets. At these events, it is common for the public to ask about the ferrets' reputed 'vicious-ness'. At these – and many other times – several ferrets are lifted from their cages and offered to members of the perpetual crowd which always gathers around such displays. As all of my ferrets are regularly handled from a very early age (some even from birth), I have no worries.

On one such occasion I placed Izzy, my favourite jill, on the top of the display while I gave a talk on the merits of the ferret (I never need an excuse for this). While I was talking, a young girl began blowing in Izzy's face and generally tormenting her. Izzy bristled and began making warning noises. I saw what was happening and interrupted my talk to ask the girl not to torment the ferrets, and then I continued with my talk. A few minutes later, Izzy was again warning the girl and I again warned her myself, as did the girl's mother. To no avail; the girl continued her tormenting and I was just about to tell the girl again, when Izzy decided that she had had enough. She uttered a loud 'chatter' and, turning around, ran across the top of the display and wormed her way into my coat pocket – her favourite travelling area and one where she knew that she would not be bothered by ill-mannered humans. Izzy, even during extreme provocation, had been the perfect lady that I always knew she was.

In normal use, a little musk is passed out with the animal's faeces, thus effectively marking the area for all of the other individuals in the vicinity. Some pet owners do not like the idea of their ferrets ever having a chance of discharging this obnoxious-smelling fluid in the home, and have their veterinary surgeon remove the anal gland, although this is unlawful in the UK. It should always be remembered that the ferret's musky odour is produced by skin secretions, with the hob's stronger-smelling urine as a contributory factor to the animal's smell, and the removal of the anal glands will not reduce these sources of odour.

Removal of the anal gland may, however, be necessary on medical grounds, and owners should be guided on this matter by their own veterinary surgeon.

*   *   *

In conclusion, I would state that the ferret is a wonderful, intelligent animal. It is eminently suitable as either a pet or a working animal and can – indeed often does – fulfil both roles at the same time. Ferrets deserve better care, and it is the intent of this book to ensure that they get this.

Chapter 2

Selection

Properly cared for – housing, diet, handling, healthcare, etc., ferrets can live for up to fourteen years, and so the choice of suitable ferrets must be given considerable thought. Indeed, much thought must be given to whether or not you are a suitable person to keep ferrets. You may consider this remark to be facetious, but I assure you that those running ferret rescue centres will not; it is a sad fact that all too many people take on the responsibility of keeping animals with little thought, and this leads to many thousands of animals being abandoned. I know of one UK-based ferret rescue society where the secretary has had over a hundred rescued ferrets in his keeping at one time, and so I urge everyone to remember that if you decide to buy a ferret today, it is going to be with you for a long time. A fifteen-year-old may not want to take his pet with him when he becomes a married man of twenty-four, but the ferret may still have almost a third of its life to live.

The fact that you are taking the time and trouble to read this book indicates that you recognise the qualities that ferrets and their relatives possess. If you are reading this book before purchasing your ferrets, you will avoid most – if not all – of the potential problems associated with keeping these highly intelligent animals. If you already have your ferrets – perhaps have been keeping and breeding ferrets for a few years – but still choose to read this or any other book on the subject, this says that you really do have the well-being of your ferrets at heart. You obviously wish to ensure that you keep up-to-date with developments in ferret husbandry, and also to see how other people look at the subject. I truly believe that no one will ever know all there is to know about ferrets, and that we can all learn from others' experiences.

Two kits from the same litter; one with eyes open, the other's just beginning to open.

Once you have given a great deal of thought as to whether or not to proceed with your decision to acquire a ferret, and you are certain that you can give the animal a real home for all of its life, there are certain other questions which you will want answering. These include:

Where and when will you purchase your ferret(s)?
What should you look for to indicate a sound, healthy ferret?
What should you avoid in your purchase?
How many ferrets should you keep and should they all be obtained at the same time and from the same person?
Which sex makes the best pet or working ferret?
How do you differentiate between the sexes?
How do you assess the character of your intended purchases?
What colour(s) should you choose?
What is the best size ferret to purchase?
At what age should your ferrets be obtained?

Let us examine all of these questions.

WHERE AND WHEN TO PURCHASE YOUR FERRET(S)

All jills (female ferrets) come into season in early spring, when the days begin to get longer, and the nights shorter. In the UK, this is in March, but in other parts of the world it will be a different month. However, some commercial breeders will alter the ferrets' photoperiod, thereby enabling the animals to breed all year round (see Chapter 6, 'Breeding' for more details). Because of this phenomenon of photoperiodism, young ferrets are normally only available at certain times of the year, i.e. summer; this then, is the best time to buy your ferrets.

At other times of the year, you may well see ferrets being advertised for sale, but unless they are bred by someone who has tweaked the animal's photoperiod, they will obviously not be kits, i.e. they will be over sixteen weeks of age. All too often, the reason for the ferrets appearing on the market is because they have proved less than suitable for their current owners. It may be that the animal has grown too big, or has not been handled properly and is therefore rather inclined to bite, or simply not tamed. If it is a working ferret, it may have developed some bad habits, such as skulking (see Chapter 4, for more information on this point). In other words, these ferrets are usually someone's rejects, and there is always a reason for that rejection. I always advise people to leave these animals alone, although I am aware that some – a very small minority in my experience – are being sold for bona fide reasons, and will make good purchases. It is far better to await the sale of kits, i.e. ferrets under sixteen weeks, although this may necessitate waiting for some months. It is worth bearing in mind the maxim 'He who acts in haste, repents at his leisure'.

Ferrets are sold by many people, from the person who gives loving care to their ferrets, and simply breeds one or two litters a year to ensure continuation of the line, to commercial breeders who have many hundreds of ferrets kept and bred in order to make money for their owners. In between these two extremes are many others, who include good and bad ferret keepers, all with their own reasons for breeding and selling the animals.

Wherever possible, go to an established ferret breeder who has a proven reputation, and who has been keeping and breeding ferrets for some years. While no such person will let you have their very best ferrets – they will want to keep those for their own purposes – neither will they risk a hard-earned reputation by selling you inferior stock.

In some countries, the only purveyors of ferrets are pet shops; as with breeders, there are good ones and bad ones, and you should take great care in choosing the pet shops with which you deal.

When buying a ferret, try to visit the breeder's home or base of operations, and look carefully at the parents of the kits being offered for sale, along with all other ferrets and/or other animals kept by the seller. All cages should be clean, of an adequate size, and not overcrowded. Look at the latrine corner; the faeces should be hard, dark brown, and have only a slight smell. Examine all of the adults: are they tame? If not, then the breeder probably has not handled them enough, and may simply be keeping them for breeding to make money.

Are they all in good condition? What are they being fed on? Many people feed their ferrets on totally unsuitable diets (see Chapter 5, for more information on adequate diets), and so the offspring of these malnourished animals may well be suffering because of this.

In the UK, it is a regular sight at the many country fairs, held extensively throughout the summer months, to see individuals walking around the showground with boxes of ferret kits for sale. These people do not have the interests of their ferrets at heart, merely wanting to get rid of the kits to save having to look after and feed them.

Never be tempted to buy a ferret from someone like this; you have never seen them before, and it is certain that you will never see them again. You have no idea of the conditions under which the ferrets have been bred, and to buy from such people merely creates a market. Many people do buy such ferrets, but many are sold to children or those who have not really thought through the pros and cons of keeping ferrets. Many of the purchasers will not have a suitable cage for their new pets, and many will change their minds before reaching home, abandoning the ferrets by the roadside – a cruel, callous and illegal act.

What to Look For and What to Avoid

A healthy ferret will have energy and a *joie de vivre,* bouncing around in its cage, and playing with its siblings. Its eyes will be bright, ears held erect, coat shiny and clean, and faeces will be hard and have only a slight smell. An ill ferret will be entirely different; lethargic, lying down and sleeping for most of the day. Its ears will be flat against the head, the coat matted and dirty, staring and greasy to the touch; it may have a discharge from eyes, ears, sex opening or anus, and it may also have diarrhoea.

Under no circumstances should you buy any ferret from a cage containing one or more ferrets as described at the end of the last paragraph; to do so would be foolish. Even if the ferret which you purchase is not exhibiting the symptoms, it is almost certain to be infected and, if you place it with others, it will pass on this ailment to all the ferrets with which it comes into contact.

HOW MANY FERRETS SHOULD BE KEPT?

There are conflicting thoughts on how many ferrets should be kept together. Some people hark back to the ferret's wild ancestor, and point out that the polecat is a solitary creature, and so the ferret is best kept on its own.

I personally believe that ferrets are happier when kept in groups (known as a 'business'). My own ferrets are kept in communal cages where I can have up to fifteen all living happily together, although this may change during the breeding season (see Chapter 6).

I really do feel that a single ferret, kept on its own all of its life, will not be as happy as it would if it had access to a fellow ferret.

Initially, I recommend that two jills are kept, as ferrets like company of their own kind and two hobs, unless kept together from a very early age may fight, especially during the breeding season. A hob and a jill kept together will inevitably produce young, bringing with them the problems of space, food and handling. A couple of years experience are needed before breeding should be embarked on.

If you are intending to hunt with your ferrets, but not to use one of the many electronic ferret detectors that are on the market, a 'liner' is essential. This is a large hob that is kept on its own and, in the event of a lay-up, is sent into the burrow attached to a long line (hence the name) to find the dead rabbit, move the errant free-working ferret and then curl up next to the dead rabbit. You then dig along the line (which is marked every metre or so) until you find the liner and rabbit (easier said than done). The liner must, of course, be big and strong enough to pull the line through all of the tunnels and then frighten away the free-working ferret. (See Chapter 7).

With the advent of good quality, efficient and relatively cheap electronic 'ferret detectors', liners have almost become a thing of the past, with more and more people turning to the new technology and ignoring the old ways. While this is, in itself, not a bad thing, it should also be remembered that even high technology equipment can break down. When this happens, it is usually at a time when it is far from convenient. At such times, it is reassuring to know that you have more than one string to your bow. The possession of, and experience with, a good liner will often prove to be of incalculable benefit to those who regularly work their ferrets.

The modern equivalent of a liner is the use of a hob with anti-social tendencies towards other ferrets (as with the original liners), but, instead of using a line, an electronic collar is fitted. Other than this one difference, the hob's use is the same as a liner on a line.

I must admit that, since I first tried one of these electronic collars, I have never worked my ferrets without one. Ferret detectors are discussed fully in Chapter 7.

SHOULD THE FERRETS BE OBTAINED AT THE SAME TIME AND FROM THE SAME SOURCE?

Some ferrets will not readily accept other ferrets, although my own experience shows that, provided the introduction is made when all of the animals concerned are young, and just after they have had a full feed (with all of the excess food removed from the cage), this is not usually a problem. True, when first introduced, the ferrets will probably have a very noisy 'fight', but this is simply to establish the hierarchy or pecking order, and usually lasts for no more than a few minutes, with no major injuries to either party. It is, however, worth checking the ferrets for wounds, and treating them accordingly, within a few hours of the fight.

If you are intending to breed from your ferrets, there are good reasons to ensure that the ferrets are not related, while it could also easily be argued that there are good reasons to ensure that the ferrets *are* related; see Chapter 10 for more details.

Related ferrets will almost certainly share similar characteristics, while totally unrelated ferrets may give birth to kits which seem to resemble neither parent.

WHICH SEX MAKES THE BEST PET OR WORKING FERRET?

Both sexes make good working animals and/or pets, but they also both have their problems, and it is worth considering these before deciding which sex to choose.

The jill (female ferret) will normally come into season (oestrus) in the spring each year and, unless mated, will stay in season until the late autumn, i.e. about six months. During this time, she will be unpredictable, bad-tempered and generally difficult to live with, qualities shared by most female mammals when they are in oestrus. It is from this characteristic, exhibited noticeably in the female dog, that we get the derogatory term 'bitch'.

A jill's vulva will swell and protrude from her body by up to two centimetres, causing it to trail on the ground, and allowing ingress of dirt and grime – potentially harmful to her. All the while she is in season, the female sex hormone, oestrogen, will be present in her body in large amounts, and this too can prove dangerous (see Chapter 6 for more details), usually leading to a form of leukaemia and thence death, if she is allowed to stay in oestrus.

Hobs (male ferrets) often become quite aggressive with other ferrets in the breeding season (although this aggression should not extend to human handlers, provided that the handling is done with care), repeatedly mating – or at least attempting to – with any females in their cage, regardless of their state of oestrus, and fighting with other males. I always recommend, therefore, that all hobs are housed separately in the summer months, in order to reduce the risks of injuries to them and other ferrets housed with them. At the very minimum, a spare cage for every hob should be available for use as and when required.

If you are certain that you will never wish to breed with your ferrets, then you should have them neutered (jills spayed and hobs castrated). If this is done, then these animals can usually live happily together throughout the year, regardless of their gender. A word of warning about castration; it is not the panacea for aggression and 'male problems' that some would have you believe, and it also carries with it certain possible medical problems (see Chapter 9 for more details).

If you feel that you do not wish to have a litter from your jills, then they should either be mated with a vasectomised hob (a hoblet), or given drugs to take them out of oestrus (see Chapter 6 for more details). Anyone who intends to take the hobby of ferret keeping seriously should have a hoblet of their own, as this is, to my mind, the simplest and most straightforward method of removing jills from oestrus and, apart from the possibility of a pseudo-pregnancy, has no deleterious side-effects. Even the short-term effects of a pseudo-pregnancy can easily be overcome (see Chapter 6).

How Do You Differentiate Between the Sexes?

The sexing of ferrets is simplicity itself, even in very young animals. In the jill, the vaginal opening and the anus are very close together, whereas in the hob, the penal opening and the anus are far apart. Do not make the mistake of thinking that the hob will have two easily visible testicles, since these do not descend until the breeding season after the birth of the hob and, even in adult hobs, they are drawn back up into the body in cold weather. Likewise, it is highly unlikely that the male's penis will be seen, as it is usually kept inside the body cavity, except during mating.

HOW TO ASSESS THE CHARACTER OF YOUR INTENDED PURCHASES

By carefully looking at and examining the parents of your intended purchases, you should gain a great insight into the characters of their offspring. Fierce, aggressive and 'untamed' parents will almost certainly produce difficult to handle kits, while 'quiet' ferrets (i.e. those which are very tame, and easy to handle) will also pass on these traits to their progeny. It is, of course, possible to tame almost any animal, but it is easier (and a lot less painful) to begin with kits from quiet parents.

When making your choice from a litter of ferrets, do not choose the ones which rush at your fingers and snap at you; these animals may turn out to be too aggressive or difficult to handle and control.

Neither should you choose the ones which hide in the background or are too frightened to come out of their nest-box, as fear can also lead to aggression. The best kits will be those that are cautious but inquisitive. Try to tempt these animals to come to you by talking gently to them, or offering small pieces of food. When they are at your hands, do not try to lift them, but merely stroke them; if they allow this to happen, even if it takes a couple of minutes, then pick them up gently. If they are unconcerned by your attentions, but inquisitive about your smell or their surroundings, and do not struggle unduly, these kits will probably grow into excellent ferrets.

Well-handled and cared for ferrets are not sly, devious or untrustworthy; neither do they bite without good reason. Since 1982, I have travelled around the UK with my Ferret Roadshow. During this time, literally thousands of people have handled my ferrets – some none too gently. Not one finger has ever been nipped – let alone bitten – during that time. This is indisputable proof – if proof be needed – that there is no such thing as a nasty ferret, only misguided and thoughtless owners.

We are all rather vain creatures, and all appreciate the opportunity to sing our own praises. If you like the ferrets that you see, tell the owner, who will undoubtedly be more than pleased at such remarks, often visibly swelling with pride. Such a compliment is worth more than gold to a true ferreter.

When it comes to time to discuss the cost, you may find that the breeder will ask only a very small sum, perhaps with the condition that, should he be so inclined, he can come to you in future years for some stock, again at a reasonable price. On the other hand, you may get a nasty surprise when the breeder asks you for several times the price that you had in mind. Do not be put off. I often use this ploy to see just how keen and committed any potential buyers are, before I will part with my precious stock. I also use the same ploy to rid myself of timewasters and people who I just do not feel would make suitable owners for my ferrets .

When you have chosen your ferret(s), take them home and place them in their pre-

Sexual dimorphism in ferrets. The hob (right) is much larger than the jill, with a heavier frame and has a broader head.

viously prepared cage, complete with water and food. Remember that these young-sters will have been used to the company and play of their siblings and, when they suddenly find themselves without this, they may be rather unsettled. For this reason, it is advisable to spend time playing with and handling them. This will also set the scene for a healthy rapport in the future.

WHAT COLOUR(S) SHOULD YOU CHOOSE?

To me, the colour of a ferret is immaterial; I want a ferret with the right characteristics for its job and for ease of handling. I have kept almost every possible colour of ferret and the one thing that I have learned is that the colour of an animal's coat does not give any guarantees about the way the animal behaves. Almost without exception, when a ferret does not behave properly, it is as a direct result of its human handler. Choose the ferret which most appeals to you.

People will give conflicting views on what makes a 'good ferret'; some will tell you that all white ferrets are susceptible to illnesses and are generally too weak. Others will say that 'poleys' (naturally coloured ferrets) are too wild and will never be completely tame. You will hear the often quoted (but still not true) view that only white ferrets should be worked, as their dark-coloured relations cannot easily be seen when working in dense cover. Equally, some people will tell you that, as most ferreting takes place in the winter months, when there is very often snow on the ground, only dark ferrets should be used, as white ferrets will not show up. As it is necessary for all of one's senses to be finely tuned and alert to all that is happening during any hunting trip, failure to see a ferret exit from a burrow should not be blamed on the ferret's colour.

I have kept albinos, polecats and almost every colour in between. They have all worked as well as I have allowed them to and the only one to make any mistakes has been me.

I like to see some variety in my stock and, provided that the ferret is a good worker, colour is not an issue. It is possible to breed ferrets of a particular colour, e.g. all pole-cats, all white, some of each and perhaps some colours which are neither one nor the other. To do this, one needs to have a working knowledge of simple Mendelian genetics and to know the breeding and parentage (for at least three generations) of all one's ferrets. Even with all of this knowledge, the actual results may never be exactly as they were worked out on paper. Full details of the genetical aspects of selec-tive colour breeding are to be found in Chapter 10.

WHAT IS THE BEST SIZE FERRET TO PURCHASE?

Size is another contentious area; some favour giant ferrets while others go for tiny 'greyhound' types. There are pros and cons to every school of thought and, if you are to remain happy with your original choice – and their future progeny – it is best that all schools of thought are examined thoroughly.

Always bear in mind that the hob will be much bigger than the jill. Hobs vary in size from about one to two kilograms, while jills can weigh as little as four hundred grams. Apart from this 'sexual dimorphism', if you are keeping the ferret for a pet, size is not really important. However, if you are keeping a ferret for working, then size is an important factor, as it is possible to have a ferret that is too big or too small for work (see Chapter 7 for more details).

Owing to their larger size, hobs tend to be worked less often than jills; this is not to say that they cannot be worked, as I have successfully used many hobs over the years. I feel that it may well be because the male does not produce young (and many ferreters wish to see fresh stock every year), that few bother to keep hobs. When the time comes for their jills to be mated, the owners simply borrow a hob from a friend. While this can be advantageous, it can also be dangerous in that, without one's own hob(s), it is almost impossible to develop a 'line' (as described in Chapter 10), the result being that litters vary in size, number of young, character and many other features. If you do decide to rely upon other owners' hobs, ensure that you check the quality and nature of the hob *before* you decide to use it; this will help ensure that you breed the type of ferret that you really want.

Ferrets come in a wide variety of shapes and sizes, all with their own adherents and champions. The arguments put forward to support each are numerous; large ferrets have more stamina (they also have more weight to carry), small ferrets are faster (but have less stamina), large ferrets are easier to see and can drag rabbits out of stops (dead ends), small ferrets can climb over such rabbits and thus flush them out of the stop. I have tried most types and am a great believer that small ferrets make the best workers. The ferrets should not be so small that they tire within five minutes of being put into a hole, but neither should they be so large that they cannot easily pass through the mesh of an average purse-net without disturbing it unduly.

Smaller ferrets seem to have the ability to climb over the backs of rabbits which have set themselves into a stop, with their nose against the wall and back feet kicking out viciously. This ability results in fewer lay-ups (i.e. killed rabbits which are then eaten by the ferret, which then decides to take forty winks). Larger ferrets will simply scratch and eat away at the back of the rabbit, until the animal is dead. This almost always results in lost meat – a waste – and digging.

Recently, many ferret breeders have been successful in producing tiny animals, in line with the legendary 'greyhound' ferrets of yesteryear. Some of these are extremely tiny – I have seen hobs no more than 200mm nose-to-tail – while others are simply smaller than normal ferrets. Recently I managed to acquire some of these animals of varying sizes and, at the time of writing, have worked them for two seasons. While the supporters of these 'pocket ferrets' will tell you that they make the best workers, my own experience says different. The ferrets which are simply smaller than normal (about 350mm nose-to-tail) are good workers, being able to manoeuvre easily in tight spaces and climb through the mesh of purse-nets without disturbing the set net. However, the really tiny ones have a *huge* disadvantage – they are so small that they tire extremely quickly and easily, and lack the strength necessary to be able to tackle awkward conies which refuse to play the game and run away from the hunting mustelid. I have had several 'backed-up' rabbits (i.e. ones which have tried to escape the ferret and reached a stop – the end of the tunnel). When a rabbit backs-up, it sticks its head down into the stop and uses its feet to discourage the ferret. A good working ferret will climb over the top of the rabbit and chase it out or, failing this, kill it where it stands. However, on several occasions when this has happened with the tiny ferrets, I have noticed that the rabbit inevitably wins, and the ferret fails to produce the goods. I can only attribute this to the ferret's lack of stamina and strength, which must go hand-in-glove with its lack of stature.

To my mind, although these pocket ferrets are cute and have lots of 'aaah' appeal as pets, they just don't cut it as workers.

AT WHAT AGE SHOULD YOUR FERRETS BE OBTAINED?

As stated earlier, I recommend that kits of between the age of eight to sixteen weeks are obtained. Both younger and older animals are often offered for sale, but animals less than eight weeks old should never be purchased, as they are not mature enough to leave their mother; like many animals, ferret kits learn much from their mother by observing her natural behaviour. Play and other activities with siblings is equally important to a kit's development.

When older animals are offered for sale, there is a good reason, and it is usually because the current owner has mistreated the animal, or the ferret has acquired bad habits. Treat with the utmost suspicion anyone trying to sell adult ferrets – even if they have a plausible excuse.

WHAT MAKES A GOOD FERRET?

Everyone has their own ideas on this subject, and very few will agree on all aspects. To me, a good ferret is one which performs its tasks well – be it a hunter or a pet – and is easy to handle, come what may. I do not like ferrets which bite, although I do realise that this trait is probably the fault of the breeder/owner rather than the ferret.

Some ferrets appeal to me because of their looks, i.e. they are aesthetically pleasing *to me*, but remember the old saying, 'Beauty is in the eyes of the beholder', and different looks appeal to different people. If *you* like the looks of a particular ferret, and the way that it behaves – its nature and behaviour – then that ferret will be a good ferret for you. If there is one aspect of the ferret that you don't like – be it size, colour, or nature – then it is a bad ferret *for you.*

Handling and Taming

Good ferrets do not bite and, to my mind, there is no such thing as a bad ferret, simply too many bad owners of ferrets. From an early age, ferret kits must be handled gently but carefully; with this type of treatment they will soon learn that your hands are wonderful, but definitely not on the menu. They will come to trust you, and look forward to your handling them. Never, during all of the time that I have been involved with ferrets, have I been seriously bitten by one of my own ferrets. On the few occasions that a ferret has nipped me, it has always been because of my own clumsiness.

I know all of my ferrets well, and they trust me implicitly. Because of this, I can handle my young stock while they are still in the nest. At this age I expect to have my fingers nibbled by the kits but, as they are so young and weak, this is not a painful experience. Nevertheless, the youngsters need to be taught that your fingers are not food, and I do this by gently tapping their noses every time that they try to bite me. They soon learn.

Throughout the time that I am handling them, I keep up a constant conversation with them, in a soft and soothing tone of voice. I believe that the voice is just as important as the hands, and can be used to calm and reassure ferrets of any age; this habit of talking to small, furry animals does, however, give rise to strange looks from non-animal people.

The next step is for me to dip my fingers into a mixture of milk and egg, offering my fingers to the youngsters to lick. If they try to bite me, they receive a gentle tap on the side of their nose; within a very short time, the kits are entirely trustworthy.

It is important to keep all movements steady and smooth; if you make sudden, unexpected movements, the ferrets may nip through fear. If that happens, it is not fair to blame the ferret.

When picking up ferrets, place one hand around their torso, with the thumb and first finger around the neck and the other hand under the backside, to take the animal's weight. To help sooth and relax the ferret, gently swing it with the hand under its front legs, and gently stroke its body by pulling it through the other hand. This method works with even the most un-cooperative of ferrets, even adults. With tamer ones, a few minutes of this treatment will induce a soporific state, rather like the 'dizzying' of a chicken or pigeon.

At first, ferrets that are not used to being handled may struggle and try to escape; at such times, do not increase the pressure on the ferret's body to try to prevent it from moving, as this may well cause injury to the ferret.

While holding it in your arms, give the animal a tit-bit, such as a small piece of food. This is a very good way to win the ferret's confidence and get it to accept you – the beginnings of a good relationship and a good rapport.

Chapter 3

Housing

One of the most important aspects of keeping ferrets, and which must be given extensive consideration and action prior to the purchase of animals, is that of housing.

In the Middle Ages, it was common for ferrets to be housed in small wooden barrels; even today, some (too many) owners keep their ferrets in cages which are either too small or of the wrong design – or both. There are two main types of ferret keepers – those who see their animals as pets, and those who see them as working animals. I accept that there are some who see their ferrets in both of these categories, but I hope that my readers will allow me this generalisation. Pet keepers tend to keep their animals indoors, while the keepers of workers tend to keep their charges outside the home. To avoid further confusion and perceived slurs, I will refer to the two categories as 'indoor ferrets' and 'outdoor ferrets'.

Indoor ferret cages ideal for breeding purposes. Cages such as this are roomy and easy to clean, and well worth the investment. Note the airtight feed container in the background, which will also protect the feed from the unwanted attentions of rodents, etc.

INDOOR FERRETS

For ferrets kept inside the home, there is little need to make elaborate and totally weather-proof cages. Indeed, some owners simply leave their ferrets free to wander around the human home, with perhaps a pet bed or nest-box in a corner, and a cat litter tray where the animals can relieve themselves.

To many indoor ferret keepers, however, this is not sufficient; there are, after all, inherent dangers in allowing ferrets the run of the house. Human feet, especially when encased in heavy shoes or boots, do nothing for the well-being of ferrets, while the animals themselves may climb into danger areas or even bite electric cables.

If your ferrets are to be allowed total free access to all parts of the human home, you must take precautions to prevent their injury. These precautions include informing all visitors that ferrets are loose in the home, and ensuring that all cupboards, drawers, etc. are kept closed. It is also advisable to keep all outside doors and windows closed too; they can act has avenues of escape for the ferrets, but may also allow access to cats or other potential predators of your ferrets.

Minimum housing for this type of ferret keeping is a cat or dog bed and a litter tray (preferably in each room but, as an absolute minimum, one on each level of the human home). Better still is to provide a secure cage into which the ferret may be fastened as and when required.

This cage need not be elaborate or even heavy duty, since it will only act as a temporary holding cage, for example when you have guests, or in hot weather and when all windows and doors are left open. Nevertheless, the cage should be secure, and have an adequate supply of bedding available to the ferrets at all times.

Outdoor Ferrets

Obviously, ferrets living out-of-doors must be supplied with a fully weather-proof cage, which is big enough, and of the correct dimensions, to allow the ferrets to indulge in their natural behaviour. It may come as a surprise to some readers, but ferrets are three-dimensional animals; that is they like to climb.

One of the comments made when people first see my ferrets in their own cages, is that they have never before seen a ferret that climbs. Without exception, this is owing to a lack of opportunity for the ferret, and not because my animals are unusual in their habits.

There are two main types of housing for ferrets; one is the 'ferret cub' (a hutch), and the other the 'ferret court' (an aviary type cage). Both have their adherents, and both have their place. My own ferrets all live in ferret courts throughout the winter, but in the summer, the hobs are all housed in separate cubs.

Whichever method of housing you choose to supply for your ferrets, you must remember that it will be home – in some cases a complete world – to the ferrets which are kept in that cage. It is obvious, therefore, that the cage should be a suitable size and shape, be easy to maintain and repair, and supply all of the items necessary for the ferrets to lead a long and happy life.

Ferret Cubs

In essence, these cages are simply well-built hutches, similar to the type that many rabbits are kept in. The minimum size for a cub suitable for two ferrets is 1.5m x 0.75m x 0.75m; it is better to err on the large size than give animals cramped quarters. A nest-box should also be supplied, and this can be fixed to the outside of the cub, thus giving more space for the ferrets to live, or may form an integral part of the cub. Access to the nest-box should be through a small 'pop-hole', measuring about five centimetres in diameter for jills, and seven to ten centimetres diameter for hobs.

Materials for the cub must be chosen with care, as they must;
– be easy to work with
– have good insulating properties
– be affordable
– be easy to maintain

A typical ferret cub, suitable for housing a ferret out of doors. (Shown without solid front door to nest-box, at left, and wire door to run.)

– be non-porous, or at least capable of being made so
– be strong and durable, giving the cage a long life.

Without a doubt, the only material which comes close to fulfilling all of these criteria is wood. In this case, the cub should be built from exterior or marine quality plywood, at least one centimetre thick and screwed rather than nailed together, to enhance the strength of the structure. Do not paint the timber as this will make the cage 'sweat', and it will never really look clean. Instead, use one of the many brands of timber treatment sold for use on garden fences and outbuildings. Ensure that the material is one hundred per cent non-toxic, and follow the manufacturer's instructions implicitly. Re-treat on a regular basis. *Never* use creosote, as this is toxic and caustic to ferrets.

The roof should be sloping to the front, where it should overhang by about five centimetres; the slope will prevent the build-up of water on the roof, while the over-hang will help prevent water from dripping into the cub via the wire front. The roof should be covered with top quality bitumen felt, or a similar material, to give complete water protection. To ensure ease of maintenance, cleaning, etc., the top should be hinged, allowing it to be fully opened for cleaning, giving unrestricted access to the cub's interior. For safety's sake, fit a device which will be able to lock the lid open; more than one ferret has met an untimely death as a result of a gust of wind blowing down the lid of the cub.

A nice effective refinement is to make the roof double-skinned, which helps insu-late the cage interior from fluctuating ambient temperatures; this is particularly important in hot weather, as ferrets cannot tolerate excessive heat. Make the gap between the skins at least 150mm, and this will act in the same way that double glazing works with windows.

If you have not given the roof a double skin, then there are temporary measures

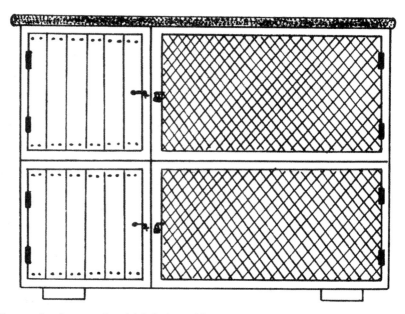

A two-tier ferret cub, which helps utilise available space to the maximum.

that you can take to cool the cub in hot weather. Place four thick pieces of timber, at least two centimetres thick, along each edge of the roof, and place a piece of sheet timber, exactly the same size as the roof, over these. Another method is to cover the roof with cloth or hessian sacking, and by keeping this material wet, the inside of the cub will remain at tolerable temperature levels. However, a little planning at the design stage of cub-making will eliminate the necessity for such actions every summer.

All joints in the cub must be tight and well-fitting, with any gaps filled to prevent draughts. The front, which should also be hinged, should be covered with top-quality weld mesh, at least 16swg and with a mesh size of either 25mm x 25mm or 25mm x 10mm. If the mesh is any larger, kits will either be able to escape, or will get stuck while trying to; any smaller, and the mesh will not allow enough light and fresh air through.

To help ensure security, fit heavy-duty hasps and staples, and use the very best padlocks; there is always someone who will covet your possessions, and it is only common sense to take precautions to help prevent the theft of your ferrets.

Raise the cub off the floor by adding legs; this will help reduce temperature loss by cold striking up from the cold, damp ground, and will also put the cub at a convenient height for you to work on, when cleaning and performing other maintenance duties.

Rather than waste the space below the cub, it is possible to turn it into an exercise area for the ferrets. This is easily achieved by covering the back and both sides with weld mesh, and putting a pair of opening doors on the front; the addition of a solid timber floor will complete the exercise area. The ferrets can gain access to this exercise area via a length of corrugated piping, such as that sold for draining soil. By providing a shutter which can seal off the access, along with a separate nest-box that can be placed in the bottom area, you will also have made a separate cub, for emergency use, or to separate ferrets in the breeding season.

I must here mention that some ferret keepers make all of the cub floor from wire mesh; I have never held with this practice, and feel that it serves no useful purpose for the ferret's welfare. The only possible benefit is to the idle keeper, who will not have to be concerned with cleaning the cub as often as it should be done; the wire floor will allow cold and draughts into the cub, thereby making the occupants suffer. If owners cannot be bothered to keep the floors of their ferrets' cubs reasonably clean and dry, then I would question whether they should be keeping ferrets at all.

To ensure that one corner of the cub does not sink into the ground, causing the cub to topple over, and that all debris from the cub is easily cleaned up, the area that the cub stands on should be of either concrete or well-laid paving slabs.

It is also worth investing in a proper path to the cub area, again made of either concrete or paving slabs; the benefits of this will be realised during bad weather, when rain and snow will turn any other type of well-used path into a sea of mud.

FERRET COURTS

If you have the space, then ferret courts are much better – for both the ferrets and yourself – than cubs, and are to be commended.

The first ferret courts were built from brick or stone, and resembled dog kennels with runs attached. I have seen a few modern courts built on the same lines, although it is true to say that these are the exception rather than the rule. Most modern ferret keepers use timber, metal and plastics in the construction of both cubs and courts.

Ferrets kept in large, outdoor courts are fitter and healthier than those kept inside or in smaller cubs (hutches).

My own courts consist of a timber frame (50mm x 50mm), covered with 5swg welded mesh. They vary in size from 4m x 4m to 2m x 1m, but all are 2m high. This height is as much convenience for myself as it is a design consideration for the ferrets; at this height, I can easily stand and work in the court, without having to adopt a very uncomfortable crouching position. This ability ensures that cleaning and maintenance are easily achieved. However, as stated earlier, ferrets, although not as agile as their cousins the martens, do like to climb; and the extra height means that the court can be furnished with branches for them to enjoy this activity, thus increasing their living space.

All of my courts have concrete floors. Some would argue that it would be more natural for the animals to be given a soil floor, complete with grass; such a floor may look nice when newly laid, but will soon become worn. In addition, ferrets also like to dig, and soil is an easy medium to dig in; escapes would be inevitable unless the welded mesh were sunk into the floor by at least 50cm. Another problem that grass and soil floors would have is one of cleanliness; it is impossible to keep such a surface clean and, when a disease occurs within the cage population, thorough disinfection of the surface is imperative to prevent the spread of the problem, but such cleaning is impossible to achieve on soil.

Concrete, on the other hand, is porous, and urine and other fluids can and do soak into the floor. It is possible to treat the concrete with paints and epoxy resins which are designed to seal the surface, making cleaning much easier. However, there have been several reports of animals kept on surfaces treated with some of these materials, which have succumbed to poisoning of various kinds. While I am sure that there are sealants which are perfectly safe for use with animals, I prefer instead to put up with the slight inconvenience of having a porous floor on my ferret courts.

In some of my ferret courts, nest-boxes are fitted inside, while on one, I have fitted all of the nest-boxes on the outside. This serves two purposes; it gives the ferrets more room in their court, and allows me to inspect and clean the nest-boxes without the necessity to empty the cage of its occupants. I can easily lock the ferrets in their nests while cleaning the run, and out of them while cleaning the nest-boxes. I can also gain access to young by placing the food for the mothers in the run, and then going out of the court to the nest-boxes and inspecting the kits; this helps alleviate stress on nursing mothers and kits alike.

There should be enough nest-boxes for all the occupants to gain benefit from them; I always have one more nest-box than I do ferrets within the cage. This enables each one to gain access, even if one or more of the other ferrets decides to start bullying. The nest-boxes should measure about 45cm x 45cm x 30cm high, and be positioned around the court, preferably at different levels, but all with the pop-holes facing away from prevailing winds. If they are raised about 150mm off the ground, this will help protect them against draughts and cold. Securely fix a piece of timber about 40mm long to the nest-box entrance, and this will ensure that the ferrets can gain easy access. These ramps should be set at about thirty degrees, and have small wooden cross-pieces, or deep saw cuts, about every 25mm along their length. This will give the animals' feet extra purchase, and enable the ferrets to use the ramps safely even when the wood is wet and slippery.

I like to have a roof which opens on these nest-boxes, in order that I can check on

When breeding ferrets, it is essential that the jill is provided with a well designed nest-box and plenty of bedding.

any litters. This gives easy access, and also helps prevent ingress of water, even in torrential rain. Place a good covering of wood shavings in the bottom of each nest-box to help absorb any water carried in on the animals' feet or fur; *never* use sawdust, as this can adversely affect the ferrets, causing breathing or eye problems. Ensure that each nest-box has an ample supply of top-quality meadow hay, with more hay needed in very cold weather, and far less in warm weather. During extremely warm weather, the ferrets may well throw out most or even all of the hay, in order to keep their sleeping quarters at the correct temperature. I am of the opinion that it is better to supply too much than not enough bedding, leaving the ferrets to make their own beds according to the prevailing weather conditions.

Although ferrets enjoy fresh air, and both a little sunshine and rain, it is essential that they also have access to shelter from these two elements, without having to seek refuge in their nest-boxes. To give the ferrets shelter, I cover about a third of the roof of the court with corrugated polycarbonate sheeting; this is translucent, allowing light through, but not creating a 'greenhouse effect' in the court. Being impervious to moisture, it also keeps rain and snow out of the court. It is essential that all sheets used have a good overlap, to prevent water seeping through the joints, and that the whole roof slopes off to an adequate gutter and downpipe, allowing water to run away safely.

The inside of the court can be supplied with various items, such as branches, plants or bushes in tubs, or even a rockery. These items will enhance the appearance of the court to human eyes, and also provide play items for the ferrets. Always ensure that any plants placed inside a court are not poisonous to the ferrets. Piles of rocks and lengths of piping (of different diameters) will help the ferrets to learn to enter such places without any fear; while this is essential for working ferrets, it is also good fun for pet ferrets.

POSITIONING

The correct positioning of your ferret cub or court is of paramount importance, and should obviously be considered before construction of the cage begins. After all, the cage's inmates cannot move from that area and are, therefore, quite literally at the mercy of the elements. Ideally, the chosen site should give protection from excesses of wind, rain, sun, heat and cold.

It is impossible to give readers precise rules on where a cub or court should be placed, as everyone's situation will be different. When making your choice, however, there are certain facts that should be very carefully considered.

Remember that in cold weather, some sunshine will be extremely useful to keep the ferrets warm, but in the summer, that same sunshine will be unwanted, and may even prove fatal to the ferrets in that cage. Bear in mind too, that the sun moves throughout the day, and take advantage of this fact. Early morning sun is much cooler than that of midday, and it is usually possible to position the cage so that the inmates benefit from an early morning sunning, allowing them to, quite literally, warm up for the day ahead. However, do not have the sun on the cage at midday, so that the animals cook in the heat.

Ascertain the direction of the prevailing winds (which will, of course, drive the rain) and ensure that there is an adequate and effective windbreak. A trellis, covered with a non-poisonous creeping plant, will make an attractive and efficient windbreak and sun shield.

CLEANING

If you are to keep ferrets successfully for a long time, it is important that you instigate a sound cleaning regime, using the correct materials and tools. As mentioned earlier, this is far easier to achieve if you plan the design properly, and give great thought to the materials used in the construction of the cage, be it a cub or a court.

Although their critics would never admit or accept the fact, ferrets are extremely clean animals; they spend much time on personal hygiene and are very particular where they empty their bladders and bowels. The latrine area chosen by ferrets is almost always in a corner, or at least against a vertical edge, and as far away from their sleeping quarters as possible. Once established, the latrine will stay in the same location for the same ferrets. Where ferrets are kept in a commune or a business, i.e. three or more ferrets, there may well be more than one latrine, but the same ferrets will always use the same latrine.

This habit makes life easy for the ferret keeper, since part of the daily routine should be to shovel and scrape out the area of the latrine(s), thus reducing the build-up of faeces, and thus smell and the risk of disease. In cubs, some owners cut a hole in the corner furthest from the nest-box, and cover it with weld mesh, of about 16swg and 10mm x 10mm. The theory is simple; the ferrets will use the area as a latrine, and the faeces will fall through the mesh into a suitable container placed under the mesh by the ferrets' owner. I have tried this, but have never witnessed much success; the faeces tend to get stuck on the wire, which is the very devil to clean. Much better, to my mind, to reinforce the area of the latrine with extra timber, plastic, stainless steel or even a few extra coats of timber treatment to protect it against the urine and faeces, and clean the area every day.

With a court the problems are fewer, provided that you have given the floor the necessary slope to allow good drainage. Clean the latrine area every day and, once a week, remove all of the ferrets from the run, and use a power jet hosepipe to clean the whole area. After being left for about twenty to thirty minutes, the run will be dry enough to allow the ferrets back in.

A good-quality disinfectant should be used to help reduce the build-up of smells and harmful bacteria, but great care is needed in the choice of disinfectant. Disinfectant agents are classed as either physical or chemical, for example steam can be used to clean cages (common in kennels and catteries), and is a physical agent, whereas chemical agents are found in the disinfectants which we can buy from many shopping outlets. All disinfectants are potentially dangerous, and yet their labelling would often lead one to believe otherwise. Some are highly toxic to certain species, while all can cause allergic reactions in certain individuals. All disinfectants work better at higher temperatures, and so should be mixed with hot water to form a solution of the correct ratio. *Never mix two disinfectants together, as the results could be highly dangerous.*

The chemicals used in disinfectants are classified into generic groups – alcohols, aldehydes, alkalis, halogens, phenolics and surface-active agents – and common disinfectants draw their constituents from the aldehydes (e.g. formalin), the halogens (e.g. sodium hypochlorite – bleach), and the phenolics (e.g. phenol). In addition, the surface-active agents are sub-divided into two more groups – quaternary ammonium compounds ('quats'), and amphoteric/polymer surface-active agents. The latter group is the one which contains the most recent additions of the disinfectants, and are both detergents and disinfectants, making them excellent for both cleaning surfaces and then rendering them 'safe'. I cannot recommend the use of these preparations enough in ferret cubs and courts.

There are two generic types of disinfectant generally available – sodium hypochlorite (bleach) and phenol.

Bleach

This is extremely effective and forms the basis of many 'cleaners' and 'disinfectants', all of which are cheap, easily available and, provided that common-sense precautions are taken, relatively safe and, therefore, useful for cleaning ferret cubs and courts. Sodium hypochlorite has a very low residual toxicity; throughout the world it is used for cleaning dairy utensils and milking machines, etc. It can, therefore, obviously be used for cleaning the ferrets' dishes and drinking bottles, provided that these are well rinsed before use.

It should be noted that all disinfectants are only effective when used on cleaned and rinsed surfaces, as even small amounts of detergent or organic material can inactivate sodium hypochlorite. Solutions of bleach and water, if left standing for any time, will also lose their efficacy, as the bleach oxidises and much is lost to the atmosphere. Because of this, it is important that all bleach solutions are used immediately they are prepared; if they are left standing, they will be ineffective against micro-organisms, and you may as well save your money and simply use a bucket of water.

After use of bleaches in ferret cubs and courts, rinse well with clean water, and then allow the cage to stand empty for about thirty minutes before returning the ferrets to their home.

Phenol

Popularly known as carbolic acid, phenol is a caustic poison which is extremely toxic to living tissue. Modern derivatives (all of which share the same properties as phenol) include xylenol, chloroxyphenol, orthophenylphenol and hexachlorophene and are equally dangerous to living tissue, being absorbed through the skin; an indication of the inherent danger of their use is that many of these substances have been banned from use in all products designed for human babies, and have long been *verboten* in catteries. When used in ferret cubs and courts, the animals may well suffer from sore feet, tails and tummies, and even 'burns' to their lips, mouths and tongues; ferrets will also die from phenol poisoning.

It is easy to tell if a disinfectant contains phenol, as it will turn white when added to water. If you care for your ferrets, DO NOT USE ANY PHENOL-BASED DISINFECTANT IN THEIR CAGE. To ignore this warning may well result in injury or death for all the ferrets in your keeping.

In the UK and most other countries, the labels on most disinfectant bottles and containers will list the chemical ingredients. However, in some countries, this is not so, and it is possible to see manufacturers claim that the substance is 'safe', 'safe in use', or 'safe when used as directed'. I have never seen any label which claims that the disinfectant is safe for use with all animals and, even if this were to be true, there could still be individual animals which have an allergy to one or more of the chemical ingredients in the product.

Many disinfectants have added 'smell', usually pine or similar; while this may make the liquid more appealing to humans, it will only mask smells. Of course, if the cub or court is properly and regularly cleaned, there will be no need for this masking. In addition, heat may well cause irritant fumes to be drawn from untreated timber into which the disinfectant has soaked.

Whichever disinfectant you use, ensure that you adhere strictly to the manufacturer's recommended dilutions and, if any animal shows signs of an allergic reaction – change the disinfectant immediately.

Cleaning Utensils

Obviously, in order to carry out the day-to-day cleaning of one's ferret cub or court – essential to avoid health problems – certain items of equipment are needed. These need not be elaborate, and most homes will already have most items 'in stock'; however, it is advisable to have a totally separate set of cleaning utensils, which are to be used exclusively for the ferret cubs and courts. In order to minimise the spread of diseases between cages, it is highly recommended that a separate set is used for each cage, and that interchange of utensils does not occur. General items that you will require are:

- Hard-bristled, long-handled broom
- Hard-bristled handbrush
- Dust pan/shovel
- Scraper (for cleaning latrine; a paint scraper is ideal)
- Buckets (stainless steel are best, although plastic will suffice)
- High-power hosepipe (or, even better, a pressure washer)
- Scrubbing brushes (deck scrubbing brushes with long handles are ideal for courts)

Pressure washers are designed to operate at either high or low pressures, using hot or cold water, or steam. They can be used with detergent, and are ideal for cleaning courts – roof, walls and floors.

Try to establish a routine, which will help you, and also ensure that the ferrets have well-cleaned cages, thus helping them lead longer and healthier – and dare I say happier – lives. To reduce the risk of smells, keep two buckets by the door of the cage, one containing a mild solution of surface action cleaner, and the other fresh water; both buckets will require cleaning out and replenishing every few days. As the dirty feeding bowls are removed from the cub or court, wash and rinse them in the buckets.

Flies are always a nuisance in warm weather, and they are best tackled by keeping a strict cleaning regime, regularly clearing the latrines. In addition, suspending fly papers around the outside of the cub, or inside a court (provided that they are well out of the reach of the inmates), will help reduce the problem. Ferrets, being natural hoarders, will hide away bits of uneaten food, often, but not always, in their nest-box. In warm weather, maggots will appear within a few hours unless the cage is kept scrupulously clean and all of the surplus food removed regularly. Even where flies are a great nuisance, one should *never* use insecticidal sprays in or around an occupied cage.

The best way to deliver fresh water to ferrets in cubs is to use a water bottle. The larger size is needed where several ferrets are kept in the same cage.

FOOD DISHES AND WATER BOTTLES

Water is essential for all animals, and you should always make provision on the outside of the cub or court to attach a water bottle. This should be of the gravity feed type, complete with a stainless steel tube and nipple, and is far preferable to the traditional dish. Dishes can easily be tipped up – even quite heavy ones – and the water may soon become fouled and undrinkable, even if the dish is emptied, cleaned and replenished daily. It is also possible for young kits to fall into bowls of water and either drown or become seriously chilled. The tube to the bottle must be checked daily, in case it has become blocked. To help inhibit the growth of algae on the inside of the bottle (which is both unsightly and a health risk), a coloured bottle, such as a wine bottle, can be used.

Chapter 4

Handling, Taming and Training

Whenever I make one of my frequent public appearances with ferrets, I always manage to amaze my audience by handling the ferrets easily – and not getting bitten. Many of the onlookers will make remarks to the effect that I must be mad, or my ferrets have had all of their teeth removed. While I cannot comment on the first accusation, I can assure my readers that none of my ferrets have had teeth removed to facilitate handling.

There is no secret behind my tame (or 'quiet') ferrets, other than the fact that they are constantly handled in the correct manner, and so neither party has anything to fear from the other. To me, this is purely common sense; I see no purpose in having an animal that one is fearful of handling, since this will mean that the animal is not handled at all, and neither will the cage be cleaned as often as it should be. All this will do is create an imprisoned animal, which will be aggressive through fear; it is cruelty to the ferret.

For some reason, there are those who deliberately create untamed and un-handleable ferrets. Their 'reasons' are either that they are to be used for hunting, or that it somehow makes the owner appear much more macho, having dominance over such a fierce animal. Neither of these are valid reasons. Ferrets used for hunting need to be just as manageable as pet ferrets; if you cannot handle a ferret in normal conditions, how will you fare when the ferret has had a fight with a rabbit or a rat, or is simply excited at the action? As for the second 'reason', I cannot justify even trying to explain the mentality of such people; all I will say is that such people should not be allowed to own any animal.

There is an old saying: 'You cannot teach an old dog new tricks'. This is not true – you can, provided that you are extremely patient and, of course, it will take more time. The same applies to ferrets. Over the years, I have had many ferrets brought to me, having been found wandering in the fields, or on roads, or even in the car parks of country fairs. Obviously, when the animals are first captured and brought to me, many of them are not handleable, as they are frightened and, in some cases, injured. However, in almost all cases, I have managed to tame the ferrets and I can honestly say that I have never been badly bitten by any ferret in my care.

I have, however, been bitten by other people's ferrets; this usually happens immediately after the owner has assured me that their ferrets are 'completely safe and totally trustworthy'. In almost every case where I have been injured, the ferrets in question exhibit clear evidence of mistreatment; some are half-starved, others have been injured in some way, while others have had some teeth removed, or even broken off. This latter practice is one which I abhor, and is still fairly common throughout the world. Even where people have the ferret's teeth removed by a veterinary surgeon,

The author's son, Tom, aged 3½ with his first rabbit, caught on Christmas Eve, 1995. Tom set his own nets, and entered his own ferret in the warren. Dad killed, skinned and gutted the coney, and Mum cooked it in a pie for supper that night. The perfect day.

Tom McKay aged 20, and still ferreting.

and under anaesthetic, I still am not happy with such an action. Where people do it themselves, often using pliers, I become very angry, and wish that the perpetrators could endure such pain themselves.

One of the old ferret books that I own dates from 1897 *(Ferrets* by Nicholas Everitt), and this shows a form of 'coping' (muzzling) which involves stitching the ferret's lips together; he states that there is no cruelty involved. It may surprise the reader to know that today, over one hundred years later, some people are still employing this method, although very few will ever openly admit to it. Even many of those who would – rightly – decry such practices as breaking or extracting teeth, or sewing a ferret's lips together as cruelty, will still employ muzzles or copes to prevent the ferrets from biting.

Why do people go to such lengths to prevent bites when the answer is so simple – do not give the ferret reason to bite. By regular, considerate handling, even the most 'aggressive' ferret will soon calm down.

I am a great believer that there is no such thing as a bad ferret, simply too many bad owners of ferrets.

Initial Handling

It is best for all concerned that handling should begin as early as possible; if you have bred a litter from your jill, the kits should be handled regularly from the time that their eyes are open and they are wandering around their cage. If the jill has been handled properly, and is truly tame, she will not usually object to this (see Chapter 6).

Begin by gently stroking the kits, at the same time talking in low, soothing tones. After only a very short time they will accept, and even welcome, this attention. After a couple of days of this, pick up the kits (one at a time) by placing one hand around its chest, and the other under its back end. This will help ensure that the animal's weight is evenly distributed, and no undue pressures are exerted on the quite fragile body.

Having established an acceptance of your handling, the next step is to get the kits to accept that your fingers are wonderful – but definitely not on the menu. To do this, mix a couple of raw eggs with some full cream milk; assuming that you are right-handed, dip the fingers of your left hand into the mix, and offer them to the kits. The young ferrets will be attracted by the smell and will start licking your fingers. Some will become very excited, and may nip your flesh; at this age, the sensation is only mildly uncomfortable. However, you must not allow them to get away with this action, as they will think it is acceptable behaviour. Instead, as a kit nibbles your finger, gently flick it on the side of the face; this should be more of a shock to the kit, than actual pain. The kit will probably hesitate, and then recommence nibbling; flick it again. After a few days of this treatment, none of the kits will even dream of biting your fingers. Hence, when the animals are strong enough to hurt you, they will have no desire to.

With adult ferrets the procedure is very similar, in that we are going to train them that certain actions are unacceptable. With rescued ferrets, which are very frightened, picking up may be quite difficult. The best way to achieve this is either with a gloved hand (a method that I never use), or by first pulling the animal firmly but slowly back-wards, by the tail. This will make the ferret automatically dig in its front feet, and try to put its weight to the front. As it does this, take hold of its torso (keeping hold of the tail) with the other hand, placing the thumb and little fingers around the ferret's chest, under the arms. Keep the other three fingers behind the head. This is the best way I have ever found of securely and safely holding a struggling adult ferret.

To calm a ferret – of any age – hold it around its torso and gently swing it back-wards and forwards, gently pulling its body through the half-closed fist of the other hand. This stroking is something that all ferrets seem to enjoy, and it always calms them.

To teach an adult ferret not to bite takes a little more time and patience than it does for a kit. Holding it as previously described, offer the knuckles of your other hand; it is important that it is the knuckles and no other part of the hand, as there should be no loose flesh for it to grab hold of (bite). As it tries to bite you, quickly move your hand away, and gently slap or flick it on the side of the face; as with the kits, your intention should be to shock rather than inflict physical pain.

Immediately after the slap, swing the ferret again, as previously described. After a few minutes of swinging, offer your knuckles again. As it tries to bite, slap it again. Repeat this process for about ten to fifteen minutes every day and, after two or three days, you should have a tame and handleable ferret.

Practice makes perfect, and I am of the opinion that no ferret can be *too* tame. I also have an allergy – to pain, which I find hurts.

WALKING ON A LEAD

Many people take their ferrets for a walk and, as with dogs, they attach a lead to the animal for safety's sake. Obviously, the ferret does not take to the lead instantly, and it is worth spending a few moments to get the ferret accustomed to the collar and lead before starting the actual walk.

The collar for the ferret should be soft but strong, is usually made from either leather or nylon, and must be a secure fit. Every collar I have bought for my ferrets has required new holes for secure fastening, and I find a special leather hole punch to be an indispensable aid for this operation. While holding the ferret securely, place the collar around its neck and mark the place where the hole needs to be. Once you have punched the hole, you can try the collar on the ferret. Ensure that it is tight enough not to come over the animal's head, while not being so tight that the ferret cannot breathe. Once fitted, place the ferret on the floor; it will immediately try its hardest to get the collar off and so it should be distracted for five to ten minutes, by which time it will have forgotten the collar, and will thus accept it.

Jill and kit in training. When using a young ferret, it helps to let it work with an experienced adult.

When this has happened, and not before, it is time to attach the lead to the collar. Again, the ferret will react to the new encumbrance, and so requires distracting until such time as it is accepted. Once this is achieved, you are ready to embark on your first walk together.

Don't be too ambitious; a thirty-mile route-march will not be a happy occurrence for the ferret. The first few walks should be taken at a very leisurely pace, and not cover more than a few hundred metres. The ferret will keep stopping at every new smell and sight, and you may well have to drag it away from some of them. Try to make the whole experience of the walk a pleasurable one, for both of you. Be patient and remember that the things which you take for granted, and to which you react in a blasé manner, will be entirely new and mysterious – maybe even frightening – to your ferret.

TRAINING THE FERRET TO COME TO YOU ON COMMAND

Another behaviour that amazes spectators at the events that I attend with my ferrets, is when the ferrets come to me on my command. For some reason, behaviour that is considered perfectly normal and commonplace in dogs, is looked on with awe when it occurs with ferrets.

All animal training is linked to conditioned reflexes; the reaction that the animal is conditioned to exhibit when it receives a certain stimulus. The great Russian psychologist, Ivan Pavlov, investigated conditioned reflexes in his famous experiments of the late nineteenth and early twentieth centuries. He noticed that he could induce the production of saliva in dogs by the sight, smell and taste of food; he then linked the appearance of food with an external stimulus – a bell ringing. After a while, he found that saliva production could be induced simply by ringing the bell.

When you go to feed your ferrets, habitually make the same noise(s). I always rattle my keys, while at the same time making a kissing noise, by pursing my lips together and sucking in through them. I do this as I come down the garden path to the courts, and continue it until I am giving the food to the ferrets, which obviously come to associate the noises with food; i.e. they are conditioned. After a few days of doing this, the ferrets will come out of their nest-boxes, and to the door of the court, on hearing one or more of the noises. Even when playing on the lawn, the ferrets will still run to me when they hear these noises.

Chapter 5

Feeding

To many people, the feeding of their ferrets is a matter that is given very little thought; as long as 'food' can be obtained at the right price, the matter is never discussed. However, the correct diet – balanced to give the right amounts of the relevant constituents – is essential for a long and healthy life.

Not so many years ago most ferret owners fed their animals on milk sops (bread and milk); some misguided or uncaring people still do so today. Speaking to these people, one hears several different 'reasons'. The first is that, if one fed meat to a ferret, it would give the animal a taste for flesh – and the animal would be more than inclined to bite the hand of anyone who came near, making the ferret totally un-tameable, or the ferret would make an extra effort to catch the rabbits it was chasing in the burrow. The second 'reason' given is that milk sops are the natural diet for ferrets. Neither argument, of course, is true.

I am amazed at how many people really do believe these theories, and how many can conjure up 'reasons' to justify the diets given to their animals. How on earth anyone can think that milk sops are natural for a ferret beggars belief; how often have you ever seen a ferret buying bottles of milk and bread in a shop – how else could they obtain such items?

Some owners give their animals a diet which they believe to be correct, without ever investigating the animal's nutritional requirements; as stated earlier, the main reason for most of these diets appears to be their low cost.

I have spent a great deal of time and effort in researching all matters pertaining to ferrets, including their nutritional requirements. I have been amazed (and worried) by the results of some of this research. There is an old saying: 'You are what you eat' and I believe that this also applies to ferrets.

Over the years I, like many others, have tried to find a good diet for my ferrets; good in that it is healthy for them, affordable for me, and readily available. I have probably tried most ferret diets except milk sops, which I have never believed in as I am only too aware of the nutritional deficiencies of such a diet. Fed a proper diet, the ferret's faeces will be very dark, solid and have little smell, and the ferret will probably live for ten or more years (I have had several ferrets which have lived to thirteen-plus years). When fed milk sops, the faeces are almost liquid, very pale in colour and extremely smelly; in human terms it is rather like living with perpetual diarrhoea – not a pleasant prospect. In addition, the ferret will lead a very unhealthy and short life (about three to five years).

A BALANCED DIET

Before going any further, it is worth considering what constitutes a 'balanced diet'. All animals require certain items in their diet, with quite large differences between species. All animals require water, fat/carbohydrate, protein, fibre, some minerals and some vitamins. A ferret fed on a truly balanced diet will lead a long, active and healthy life, more than repaying you for the cost of its diet. The main components of any diet are proteins, vitamins, minerals, carbohydrates, fats and fibre.

Protein

Proteins, or amino acids, are essential for growth and tissue maintenance, and should form at least thirty per cent of the ferrets' diet. They are present in meat, eggs and milk. There are many different proteins, all consisting of different arrangements of about twenty amino acids, but it is not necessary to differentiate between them here.

It should be obvious that while all ferrets require a high-protein diet, it is even more important when considering nursing mothers, their kits and young ferrets in general. Likewise, males used for stud also require a higher protein content in their diet.

Carbohydrates

Carbohydrates provide the body with heat and material for growth. As they are made up of carbon, hydrogen and oxygen (which combine to form cellulose, starch and sugar), excess amounts are stored as fat in the body, often leading to obesity. As everyone knows, this can cause medical problems and difficulties in the ferrets' breeding; a fat, overweight ferret will not be as active, nor as agile as it should be, and so will not be as happy.

Vitamins

Vitamins are chemical compounds, essential for growth, health, normal metabolism and general physical well-being. Many vitamins play an important part in completing essential chemical reactions in the body, forming parts of enzymes – chemical catalysts. Some vitamins form parts of hormones, which are the chemical substances that promote the health of the body and reproduction.

There are two main types of vitamin – water-soluble and fat-soluble.

Water-soluble vitamins cannot be stored in the body, and so the day's food must contain the day's requirements of these vitamins. Fat-soluble vitamins can be stored in the body and, if too many are taken in at one time, they can be stored for use when the body needs them. However, an excess of such vitamins may cause toxic levels to accumulate in storage areas such as the liver. It is important to remember that an excess of any fat-soluble vitamin can lead to long-term physical problems. You should also remember that a lack of essential vitamins can be detrimental to the ferret's health.

The ferret requires:

Water – essential for all life
Protein (30%) – for tissue and muscle growth and repair
Fat (15 – 40 %) – for heating the body, and providing energy
Calcium (1%) – for teeth and bones
Phosphorous (1%) – for teeth and bones

Salt (1%) – a vital mineral which acts as an electrolyte regulating the balance of water inside and outside cells
Fibre (20–25%) – as an aid to digestion
Vitamin E (about 250 mg/kg) – promotes normal growth and development, acts as an anti-blood-clotting agent, and promotes normal red blood cell production.

Vitamin E is found in the yolk of eggs and in certain vegetable oils. A lack of this vitamin can cause infertility, heart and circulation problems and skin complaints.

Fat provides the ferret's body with energy but, given too much, the body will store it in the tissue. This can lead to many problems, including mating difficulties, heart problems and the reabsorbtion of foetuses.

Fibre (or roughage, as it used to be called) is essential for the well-being of the ferret's digestive system. Fibre will keep the ferret 'regular' and this will help prevent many of the diseases – including cancer of the bowel – with which some ferrets are afflicted. Fibre for the ferret is to be found in the fur and feathers of its prey.

The mineral calcium is found in liver, milk and milk products, eggshells, fish and snails. It is also to be found in the bones and teeth of all animals. Phosphorous is found in liver, milk and milk products, and fish.

PROVIDING THE FOOD

Ferrets are carnivores, a word derived from the Latin *carnis,* meaning flesh, and *vorare,* to devour. As such, their natural diet is one of flesh, i.e. whole carcasses of mammals, birds, reptiles, fish, amphibia and even some invertebrates. While complete diets have much to offer, there are still some ferreters who stick to a more traditional diet for their animals and feed complete cadavers of mammals and birds. For this diet to be efficient, it must consist of a wide range of species, and the cadavers should be offered with the major organs (liver, heart, lungs and kidneys) intact. The removal of the stomach and intestines is recommended, as these consist merely of thin skin, filled with waste material, and will serve only to soil the ferret's cage, attract flies, and produce unpleasant smells. All cadavers given, however, should be left in their skin/feathers, as this provides the ferret with essential fibre.

Do not, however, throw a complete cadaver into the ferrets' cage and leave it until it is all eaten. Such practices, which are all too common, are bad husbandry and will, eventually, lead to illness and possibly the death of one or more of your ferrets. Unless you keep a reasonable number of ferrets in one court, rabbits should be cut into portions of various sizes, all with the fur left on. Leave the rabbit's organs inside, but check for such diseases as liver fluke (visible as white dots on the infected animal's liver), rejecting any such organs. If you have six or more ferrets in a court or a litter of kits which are eating flesh, a full rabbit, minus its guts, will be perfectly acceptable and will help the development of the kits.

Unless one has the opportunity and good fortune to have access to large areas of land – and the animals and birds that live there – this kind of diet can seem impossible to supply. Do not despair. There are acceptable alternatives.

Road casualties are a good source of food for ferrets; I have even seen feral ferrets and wild polecats eating such animals. On journeys through the countryside of most western countries, an alert driver will spot many dead animals by the roadside. These are usually rabbits, pigeons or similar species, with the occasional songbird and rodent. Wild rodents should not be fed to ferrets unless you can be reasonably sure that they are free from disease and have not been poisoned. Even then, the cadavers should be thoroughly gutted and inspected before feeding, but I would not recommend their feeding to ferrets, as the risks of infection and poisoning are so great.

Never feed 'old' meat from bodies that have been on the roadside for more than a few hours, as the flesh of these animals or birds may already be infected with the *Clostridium botulinum* bacteria, the cause of botulism, a disease which is almost always fatal to ferrets. Always check the cadaver of any road casualty for signs of illness – puffy eyes, festering sores and/or poor fur, skin or feathers – and discard all that are suspect.

Many road casualties are young animals, which simply have not had the chance to learn of the dangers of the world and these, of course, are ideal to feed to the ferrets.

Friends and farmers who shoot can often be relied upon to provide the cadavers of pest species, or even game species which are too badly shot for human consumption or have been badly chewed by a hard-mouthed dog. If you use your ferrets for hunting, then your own catch can also be utilised. There will, however, come times when this is still not enough to provide all of the food for your ferrets; a regular supply of good-quality food is required for the animals' well-being, and your peace of mind.

Day-old Cockerels

One of the items which I used to believe was good for ferrets was day-old cockerels (chicks). These are widely available, as they are the culled male birds from hatchings of eggs to supply new stock to egg producers. They are sold mainly as food for raptors – owls, hawks, etc. – and their cost is extremely low. A fully grown hob will eat two or three of these dead chicks each day, with a jill obviously eating less. However, there are great risks to ferrets fed on a diet in which day-olds feature greatly.

Research indicates that day-old chicks are very low in calcium, protein and fat, and contain very little vitamin E. The feeding of day-old chicks has also been shown to cause such problems as hypocalcaemia, actinomycosis (a thickening and swelling of the neck), osteodystrophy, thiamine deficiency, posterior paralysis (the 'staggers'), and other maladies (see Chapter 9).

Prior to my research on the subject, I was feeding my ferrets with day-olds about every other day. After my research, I changed this, and now do not feed these items to my ferrets.

Dog and Cat Foods

A ready supply of quality meat is always available from supermarkets, in the shape of tinned dog and cat food, and many ferreters use this as a basis for their ferrets' diet. However, these are obviously not designed for ferrets, and many will need to have vitamin and mineral supplements added before they can be fed. Your veterinary surgeon will be able to offer you sound advice as well as providing the necessary supplements, if they are required.

Many pet shops now sell 'pet meat' or brawn, of different flavours, and some of these brands can be fed, but again, many will require supplementing to ensure no long-term ill-effects if these items are to be given to your ferrets regularly or over a long period.

At the very least, I recommend that all tinned dog and cat foods (and other similar foods) be supplemented with the addition of bonemeal powder to ensure that the ferrets receive adequate amounts of calcium, vital to the well-being of the ferrets, as it is essential for the animal to grow strong bones and teeth.

Dry Foods

The modern trend in animal nutrition is to feed 'complete diets', and in zoos, where the idea started, and for dogs and cats, these are now the most popular types of feeding. The foods are usually made in pellet or 'muesli-type' form. Initially, the idea was driven by necessity, as zoos needed to be able to supply a wide range of animals with all of their nutritional requirements, and found it almost impossible to do so by feeding 'natural' foods, many of which were just not available locally, or on a regular basis. The idea was highly successful and it was not long before the pet trade took an interest in the idea.

At first, the idea of feeding one's dog or cat with pellets or 'porridge' caused many people to avoid such foods but, eventually, the public discovered for themselves the merits of such diets. They are very convenient, have very little smell, are acceptable to all but the most pampered dog or cat and really are a 'complete diet'. Add to that cheapness, ease of availability and storage, and the fact that tinned pet foods contain a lot of water, and it is obvious why they have become so successful. However, unless you obtain complete ferret diets, i.e. those designed specifically for the ferret, you will have to add supplements and be ever-watchful for long-term deleterious effects on the ferrets. Such effects could be so insidious that they are difficult to notice until it is too late, by which time complete kennels of ferrets could have been made useless.

A warning of embarking on untested diets was given by BP Nutrition (UK) Ltd, at a symposium organised by the Association of British Wild Animal Keepers (Management of Canids and Mustelids, 1980). In the paper presented to the meeting, the story was told of mink farmers using a commercial diet for their animals which had not been proven with mink. At first, everyone was happy with the results but, after more than eighteen months of being fed on this diet, the mink began to exhibit serious problems, including stunted growth, poor fur, and reduced litter sizes. We should learn the lesson and not feed items or diets which are not intended for our ferrets.

When I wrote the first edition of this book, complete (dry) food for ferrets was available in the USA, but not in the UK. However, shortly afterwards, I was approached by a leading British pet food manufacturer – James Wellbeloved. I had already had experience of this company, having fed some of their products to my working dogs (English springer spaniels and Jack Russell terriers) for several years, and knew that they produced a top-class product, and so I agreed to meet them to discuss the possibility of producing a complete balanced diet, in dry pellet form, for ferrets. I had previous knowledge of such feeds through contact with friends in the United States of America, where complete ferret feed was the norm, and knew of the possible advantages of such a product – although at that time they were not, and never had been, readily available in the UK.

Wellbeloved had already produced a food which met the nutritional criteria for ferrets, and I began feeding trials, dividing my ferrets into three distinct groups (with over a hundred ferrets this meant we would still have a reasonable group size on which to base our results).

The groups were:
1. Fed solely on James Wellbeloved Ferret Complete.
2. Fed on a normal diet of small animal carcasses.
3. Fed on a mixture of small animal carcasses and James Wellbeloved Ferret Complete.

These tests were originally conducted for twelve months, but were extended for much longer.

The results were extremely encouraging – all groups were fit and healthy, well able to do the work expected of them, and the animals used for breeding produced litters of good, strong kits which went on to become fit and healthy adults. The animals' coats were thick and shiny and all had a look of total fitness.

The only real differences among the ferrets were the faeces they produced and the numbers of flies attracted to those faeces. Ferrets fed solely on James Wellbeloved Ferret Complete produced fewer faeces which were more solid than those produced by the ferrets in the other groups. As a result of this, fewer flies were in the cages of the first group than were in the other two groups' living quarters. Also, because there was no risk of the food deteriorating in the heat, the dry food could be given at any time of day without the inherent risks of food poisoning associated with feeding fresh meat. The one downside, especially for those showing/exhibiting their animals, was that the dry food stained the ferrets' teeth.

James Wellbeloved Complete Ferret Food was the first complete dry ferret food available in the UK.

As James Wellbeloved Ferret Complete became more and more popular, other pet food manufacturers began producing complete dry diets for ferrets and more and more ferret keepers turned to this feeding regime.

It is not overstating the case to say that this new feeding regime has been revolutionary and has had a huge impact on ferret keeping in the UK. Many more people, who would once have avoided the animal because of its penchant for eating meat, etc., are now keeping ferrets as pets.

Almost every ferret keeper now appreciates the convenience of being able to obtain feed for their charges easily throughout the year, and the ability to store the feed easily without the need for refrigerators and freezers. Even many ardent rabbiters feed their animals dry food during the summer when rabbiting has to cease, causing the normal food supply to dry up.

Since 1995, we have fed our ferrets solely on James Wellbeloved Ferret Complete and feel that this is the best diet for our hard-working mustelids.

One small and obviously unintentional advantage of feeding only dry food to ferrets is that the ferrets will not eat any rabbits that they kill during their work

Complete diets for ferrets have revolutionised ferret keeping in the UK.

underground, thus saving us much time and effort digging out a fed-up ferret which has decided to eat its quarry and sleep off the excess.

Ferrets, like many other animals, develop feeding tastes and preferences at an early age, i.e. they develop a liking for the foods eaten when weaning while refusing foods/tastes which are new to them. These taste preferences stay with them throughout their life, a process known as 'olfactory imprinting'. This occurs when the exposure of animals to olfactory cues during specific and restricted time windows leaves a permanent memory (the 'olfactory imprint') which shapes the animal's behaviour upon encountering the olfactory cues at later times. Thus ferrets, fed all their life solely on a complete pelleted diet, refuse to eat meat, even when this is freshly killed rabbit.

A word of warning – a food which is a complete balanced diet will not be if you add things (other than water) to it – e.g. by adding meat, you will put the feed out of balance. While an occasional treat will not be injurious to your ferrets, it is not necessary and may just have the opposite effect from that intended.

Chicken

Many years ago, I moved home to a small village in a new area, and found myself in the position of having over a dozen ferrets and no ferreting. Not only was I deprived of my favourite sport, I also had to find an alternative supply of food for the ferrets. As it was spring and my jills were pregnant, I had to find a good, reliable source quickly; young ferrets seem to have an unseemly appetite and this acted as a spur to my efforts.

I was shopping in the local village when I discovered a delicatessen which sold roasted whole chickens. Asking to speak to the shop's manager, I soon discovered that all of the chickens arrived at the shop complete with their giblets, which were then removed and discarded as waste material. I saw the opportunity and grabbed it with both hands, offering to collect all of these unwanted giblets on a regular basis; the manager also felt that this was his lucky day and he eagerly accepted my offer, telling me that I was doing him a great favour. I later discovered that the shop had received several complaints about smells caused by the decomposing giblets, especially during the warm weather. To ensure that no one was ever left in the lurch, I always collected the giblets on a regular basis, even though on some occasions I did not need them; at such times they were simply burned in the garden incinerator.

My work has forced me to live in many parts of the UK and I frequently encounter problems acquiring a good, reliable source of food for my ferrets. In 1990 I moved to Derbyshire and once again had to think about food for my business of ferrets. As I live in the middle of farming country, I checked local business directories for farms which specialise in the production of eggs. It was not long before I found one who was willing – and even keen – to help me. As the reader is probably aware, chickens only have a limited life for laying eggs at the frequency and quality demanded by today's farmers, who are businessmen. Once their useful life is over, they are disposed of. You may or may not agree with this course of action, but it is reality, and I believe in making the best of all situations. After arranging to meet the farmer, we came to an agreement whereby, when he had a certain number of chickens which had reached the end of their working life, he would sell them to me for, quite literally, pennies.

Opportunities like these exist throughout the world and I can recommend that they are taken; however, it is important that, once you have made an agreement you abide by it, or you will find yourself having to find another source of ferret food.

Beware of feeding capons to your ferrets; these are birds which are chemically sterilised and, by feeding such birds (or parts of them) to your ferrets, you may cause all kinds of problems. There is some evidence that the hormones contained in these birds can cause congenital deformities in ferrets.

Poultry in general is easily infected with harmful bacteria and is often traced as the source of outbreaks of food poisoning; take care when feeding it to your ferrets. If chicken, or any other meat for that matter, has already been frozen but has now defrosted, never re-freeze it, as this will almost certainly cause health problems to the animals to which this re-frozen meat is fed. All frozen meat must be thoroughly defrosted before feeding, and then it must be fed immediately, to help to ensure that such ailments as botulism do not affect your stock. If you are not sure about any meat, boiling it for about fifteen minutes before feeding it to your ferrets will kill harmful bacteria.

Other Foods

Raw green tripe, the unprocessed stomachs of cows, is an excellent food for dogs, cats, ferrets and almost every other carnivore. In some countries, it is easily obtainable, with some pet shops selling it in 'tubes', minced, frozen and even in tins. However, since the BSE ('mad cow') outbreaks in the UK, it has become very difficult to obtain large quantities of green tripe directly from abattoirs.

Tripe also has a very strong and disagreeable smell; many people will not feed it, solely for this reason. Ferrets, however, are not deterred by the smell, and will eat it with gusto. As with all meat, always ensure the tripe is fresh and, if frozen, defrost it thoroughly, feeding it as soon afterwards as possible.

Heart, lights (lungs), cheek and udder can be bought reasonably cheaply and make good ferret food, but again availability is limited after the BSE scares. Liver is enjoyed by all ferrets and has much nutritional value; however, it should not be fed in excess and must *never* form a staple. Avoid liver, often sold as 'pet food', which has been branded as 'unfit for human consumption', for whatever reason. My personal maxim is that if meat is not fit for me to eat, it is not fit for my ferrets either.

Many butchers, supermarkets, pet stores and other retailers sell 'minced pet food'. This usually consists of many different sorts of meat minced up together and frozen, and will often include offal, fat and waste. It is an open invitation for harmful bacteria; therefore it must be fed immediately it has been properly defrosted. If in any doubt as to its suitability, boil it for at least fifteen minutes before feeding it. These meats also tend to be high in fats and should not be fed as a staple, although the high fat content will be useful in cold weather or when the ferrets are working hard, as their calorific needs will be much higher than normal.

Fish is relished by ferrets but should only be fed in limited quantities; avoid smoked, salty or fatty fish. All fish must be well filleted, as fish bones can easily become lodged in a ferret's windpipe, often with fatal results.

Eggs

In the wild polecats eat eggs, and so some ferret keepers argue that these must, therefore, be good for ferrets as they are 'natural'. They then proceed to feed their ferrets with a regular diet of eggs, sometimes giving each ferret a full egg every day. What these people do not realise is that, in the wild, birds only lay their eggs at one time of the year – spring. It is only when birds are brought into captivity and their photoperiod manipulated, that they will supply eggs throughout the year.

Too many raw eggs (i.e. more than one per week for an adult ferret) will have a bad effect on ferrets; they can cause diarrhoea and even hair loss. Try feeding hard-boiled eggs, complete with their shells, which are a good source of calcium, or even scrambled eggs. Ferrets love them.

Adjusting the Diet

Don't forget that the animal's diet will have to be adjusted to compensate for its lifestyle; a ferret kept indoors and not given much exercise will need far less energy food (i.e. fat) than an animal kept outside and worked regularly. All animals will need more fat in colder weather. Pregnant jills and young kits will require higher amounts of protein and calcium.

How much food should a ferret be given each day? This is a very difficult question to answer, as it will vary from season to season, and from day to day, depending on ambient temperatures, energy expended by the ferrets, and many other factors. A general rule of thumb is to feed the ferrets with an amount which you think is correct, and then check on the food about an hour later. If all of the food has been eaten, always remembering that ferrets are hoarders, and some may have been hidden away in the nest-box or elsewhere, feed a little extra next time. Ideally, there should still be a small amount of food left about an hour after putting the food in the cage. Some people decry this as wasteful, but I prefer to ensure that all of my ferrets have sufficient food for their needs, rather than try to save a few pennies. All left-overs must be removed at the end of each day.

As the ferret's digestive system, like that of all carnivores, is quite short, too large a quantity of food will result in much of it passing undigested through the ferret. This is one reason why it is not satisfactory to feed the ferrets with one large meal every couple of days. Ferrets, like most animals, appreciate regular meals, and this will result in less waste; less opportunity for the food to go off before the ferrets have a chance to eat it; less illness amongst the ferrets, and happy and contented animals.

When feeding dry food, ferrets should have ad lib access to the dish of food. The dish should be emptied and cleaned before refilling.

WATER

Clean, fresh water is essential for all animal life, and so must always be available to your ferrets. Water is even more important to ferrets fed on a dry or semi-moist diet. It is best delivered in a drinking bottle with a stainless steel spout; many pet shops sell one-litre sized 'rabbit drinking bottles', which are ideal. The bottles should be checked daily, ensuring that the water flows freely, as the spout can easily become blocked with wood shavings, hay and even fur. Ensure that the bottle is securely fastened to the cage, on the outside. Dark-coloured glass, such as wine bottles, are a good method of slowing down the growth of algae on the inside of the bottle.

Regular checks must be made on all water bottles, especially in extremes of weather. During periods of very low temperatures the water in the bottles may well freeze, and often this will cause a glass or plastic bottle to crack. Covering the bottles with insulating material such as is designed for water pipes is a good way of preventing this; the one drawback with this is that it is impossible to check the contents of the bottle without physically removing the covering every time. This is, however, a small price to pay for the peace of mind of knowing that the water is unlikely to freeze.

In courts, where there are several ferrets living together, it is essential to have two or three bottles and maybe even a water container such as is given to chickens.

*   *   *

A well thought-out, properly balanced diet will help keep your ferrets free from illness and disease. While this is obviously good for the ferret, it is also good for you, as it will mean less hassle and fewer vet's bills.

Chapter 6

Breeding

In some countries, almost all ferrets sold to members of the public have been neutered. Some suppliers claim that this is done to help the owners and prevent certain illnesses related to females which are not mated every year; some even claim that this practice reduces or eliminates smells and removes all aggressive instincts from the ferrets, particularly the hob. However, the main reason for the practice seems to me to be to prevent the ferrets being used for breeding.

For the rest of us, however, breeding is essential if we are to develop our idea of the perfect ferret, thus producing our own blood lines, as do breeders of dogs, cats and racehorses.

WHY BREED?

Before embarking on any project, it is always wise to ask yourself why you are doing it and what you wish to achieve; the breeding of any animal is a project which requires long and careful consideration, and must always be carried out for a valid reason. For most of us, that valid reason is the continuing quest for our elusive goal – the perfect ferret.

When I began with ferrets, I was lucky in acquiring a trio (one male and two females) which were of good breeding and looks. However, within a couple of years, I was sure that there was scope (and to my mind a need) for improvement. I was using the ferrets for rabbiting and had encountered a few problems which I associated with the size of the ferrets I was using; I considered them to be too large. My theory was – and still is – that if a ferret is too big, it cannot move quickly enough down the long and winding tunnels of a burrow. Worse still, if the rabbit runs into a 'stop' (a cul-de-sac), a large ferret can do nothing other than scratch away at the rabbit until the luckless bunny is dead, by which time the ferret is tired and hungry, and will eat its fill before curling up and going to sleep. Conversely, a ferret that is too small may well not have the physical resources to work for long enough periods to be of use, and may even be dragged around the burrow by a large rabbit that it manages to sink its teeth into.

I decided that I would produce my own 'line' of perfect rabbiting ferrets.

You may not wish to have such lofty and egotistical ideas, but simply want to have more ferrets like the ones which you now own, or you may wish to increase the number of ferrets and don't want to buy any. Do not, however, make the mistake of believing that you can make money out of ferret breeding; unless you have huge numbers, and operate a proper business, you will never make money from such an undertaking.

RECORD-KEEPING

To my mind, one of the most vital items that you must have for any breeding pro-gramme is a good record-keeping system; by 'good', I mean accurate, up-to-date, efficient and effective. Such records are indispensable for fixing a new mutation or for selecting suitable breeding stock. By using the records carefully, it is possible to select stock that will produce good-quality kits, or always produce large (or always small) litters. It is a sad but true fact that very few breeders of small animals keep any records at all, and the number who keep *good* records is tiny.

There are various methods which you can utilise, from note-books to card indexes, loose-leaf pads to personal computers. I am well known as a 'gadget man', but I will only continue to use such items if they really are useful to me. The most valuable asset that any of us possesses is time; if by using a 'gadget' I can save myself some time, I feel that I should do so. With all new equipment and procedures, however, one must remember that there is always a learning curve – often quite steep – and so it will take time and effort before one is fully conversant with the equipment or procedure; only when you are fully conversant with it will you begin to reap any benefits.

It is imperative, if you are to be successful in your ferret breeding, that you main-tain complete and accurate records from day one. These records can be kept on a card affixed to each cage, and/or in a book or some other central device. Whatever method you use, the record should contain the following information;

> Variety
> Name and/or reference number
> Breeder, if applicable
> Date of birth
> Parents, grandparents, and great-grandparents
> Siblings
> Breeding/mating details (including sizes of litters)

If you attach cards to the ferret cages, you will be able to see at a glance every relevant detail about the cage's inmates. All writing on these cards should be done using an indelible (waterproof) pen, since other inks will run when damp or when water is spilled on them. Never use pencil, as the writing will fade and you will have lost all of the information. Never trust to memory either, as it can fail all too often.

Not all of the above details need be written on the card on the ferret's cage, but all (and more) should be recorded in the 'stud-book', a central record (not necessarily a book) of all details pertaining to your ferrets. Without such a record, you will never to able to breed your animals successfully in an ordered and efficient manner, neither will you be able to perpetuate a mutant colour, thereby risking losing it for ever.

It is a matter of convention that all males are given letters of the alphabet for recording purposes, while females are given numbers. This is because, on average, more jills than hobs will be kept. Some breeders combine these codes to identify any litters produced from a mating. For instance, if ferret 'A' (male) is mated with ferret '1' (female), and this mating produces a litter of two males and one female, the males will be coded 'A1A' and 'A1B', and the female '1A1'. In the case of the male 'A1A', the A1 denotes that the parents are A and 1, and that the specific animal is a male, which is given a letter of the alphabet as its reference; the second male from the same litter will be AIB. I find this method too complicated and fussy to maintain. Instead, I simply number all my females from ' 1 ' to infinity, and the males ' A' to 'Z'; when I get to the end of the alphabet for the first time, I start again, prefixing all letters with the first letter of the alphabet, e.g. 'AA', 'AB', 'AC', 'AD', etc. When I get to the end of the alphabet for the second time, I prefix the letter with the second letter in the alphabet, e.g. 'BA', 'BB', 'BC', 'BD', etc. When I come to the end of the alphabet for the 26th time, I prefix the letter with two 'A's, e.g. 'AAA', 'AAB','AAC','AAD',etc.

THE STUD-BOOK

Many breeders use a loose-leaf folder for their stud-book, and are quite happy with this arrangement. However, when using such books, it is very easy for a page to be lost. A hard-backed note-book of A4 size is ideal, and you will have no trouble with pages going missing. Either enter the details of the females at one end of the book, and those of the males at the other end, or use two separate books. A typical page may be set out as follows

NAME: REFERENCE:

VARIETY:

DATE OF BIRTH:

BREEDER:

PARENTS:

PATERNAL G. PARENTS: MATERNAL G. PARENTS:

PATERNAL G.G. PARENTS: MATERNAL G.G. PARENTS:

SIBLINGS:

NOTES:

MATING AND BREEDING RECORD:

Computers

Most households now have a personal computer (PC), and this can easily be used to store records, indeed the main advantage with using PCs for record-keeping is the ease with which relevant information can be retrieved and displayed. These computers may be in the form of desk-top, lap-top, note-book or even pocket-sized; the last are often referred to as palm-tops or organisers. I find that a palm-top, which can easily fit into a pocket, is extremely useful for recording details while one is actually with the ferrets. With suitable software, this information can be 'uploaded' onto the PC back in the warmth of the home. Of course the PC itself is useless without the appropriate software, and so the choice of this software must be given thought.

For record-keeping, some form of database is needed. This can, if the operator understands the system thoroughly, be very complicated and give many facilities which can prove invaluable. On the other hand, it is quite possible to buy software which is perfectly adequate, and which is user-friendly.

All records kept on computer have the advantage of being easily accessible, provided that some thought is given to their format, the fields chosen, and their storage. It is essential that you 'back-up' your information after every use of the PC. There is a risk that a hard disk will 'crash' and lose all of the information stored on it. While this is certainly not an everyday occurrence (thank goodness), it does occasionally happen and, if you have not backed-up your files, then you will be in a sorry state. As stated earlier, if you are storing data solely on your hard disk, such a crash may cause you to lose all of the information that you have gathered over many years, and so backing-up should be a regular habit and carried out *every* time that you use your computer.

LINE-BREEDING AND INBREEDING

Inbreeding is the mating together of closely related animals, e.g. mother to son, father to daughter, brother to sister. There is a commonly held belief that inbreeding is highly undesirable, resulting in poor stock and small litters. This is not true. *Controlled* inbreeding is the method that has been used for countless generations to produce champion racehorses, cattle and dogs. If inbreeding was deleterious *per se,* then we would have no Syrian ('golden') hamsters in captivity, as the original stock from which the vast majority of today's hamsters have descended came from one female and her litter found in Syria in 1930. The only dangers with inbreeding occur when there is little or no control, or when not enough animals are available to allow selective breeding. *Controlled* inbreeding will fix good characteristics, and bring out others which are not readily apparent.

Line-breeding is a less acute form of inbreeding, in which more distantly related animals are mated, e.g. cousins, grandparents to grandchildren, etc.

New colour mutations are more likely to appear through inbreeding, as there is more likelihood that both animals in a mating will possess the same recessive mutant gene (see Chapter 10 for more details).

In practice, most breeders use a combination of the two methods, keeping their own animals which are usually bred together (i.e. inbreeding), and occasionally borrowing a stud hob from another breeder – an 'outcross'. This is an animal that is entirely unrelated to its mate, and so will possess totally different genes; the introduction of unrelated blood is said to supply 'hybrid vigour'. According to some adherents of the theory, hybrid vigour will result in the offspring being tougher, more fertile, able to produce bigger and better litters, be less susceptible to illness and disease, and generally be fitter for survival. I feel that the idea has become overstated by many and is not the panacea that some would have us believe. The use of a 'shared' hob holds many dangers, particularly from the spread of disease.

The big danger with an outcross is that the animal could introduce genes which, although not showing any deleterious effect on that animal, when combined with the genes of other ferrets could prove disastrous. Again *control* is needed in all breeding programmes, if success is to be achieved.

SELECTION OF BREEDING STOCK

There is an old stockman's saying about breeding animals that is as true for ferrets as it is for cows, horses, dogs or any other type of animal: 'Put the best to the best, and hope for the best.'

For your breeding to be successful, you must start your programme with the best animals that you can obtain. If you do not own animals of the best calibre, then you should seriously consider buying young stock, rather than breed your own animals from mediocre stock since, to coin another adage: 'You will only get out what you put in'. In other words, poor stock bred with other poor stock will only produce more poor stock.

A halfway house between breeding ferrets from your own stock and buying animals that another breeder has bred, is to borrow a top-grade male to mate with one of your females (the male in a mating is known as the 'sire', while the female is known as the 'dam'). Your female must, of course, be of a high standard, or you will simply be wasting your time. You will probably have to pay some price for the stud services of the male (although some breeders will give you this service free, simply because they value the ferret fancy and wish to encourage newcomers). The payment for these services can be cash, but is often 'pick of the litter', which can mean up to fifty per cent of the litter. Unfortunately, if the supplier of the sire wishes to obtain good stock (just like you do), he or she will almost certainly choose the best of the litter, leaving you with the rest. For this reason, most breeders will supply this service free of charge on the understanding that, at a later unspecified date, if they need a similar service or some new young stock, you will provide these at the same rates – free. Note the earlier warning about the use of shared hobs and out-crosses.

If you do not know whether the animals that you possess are good, bad or indifferent, then you should ask an experienced breeder for an opinion. Failing that, if you intend using your ferrets in exhibitions, carefully read the standards set out by the various ferret clubs, and compare your ferrets to these standards. However, it has to be said that reading about a subject is not as good as having first-hand experience of it. Attendance at a show will give you the opportunity of talking to experienced breeder/exhibitors about your stock and what makes a good ferret. Remember that each different variety will have a different standard for its colour and markings, but all ferrets have the same standards relating to size, type, condition, eyes and ears. Remember also that different clubs and different judges will interpret the standards slightly differently from each other, and so you should try to visit shows in other areas, rather than merely staying in your own locale.

BREEDING SEASON

For readers who have a serious interest in breeding their ferrets long-term to achieve a good working line of animals, I would recommend reading a copy of my book *Ferret Breeding* (Swan Hill Press, 2006), as this gives precise details of how to establish and implement a breeding programme to ensure the best results.

The ferret's breeding is controlled by a phenomenon known as photoperiodism; in essence this means that the animal's brain decides when the breeding season is here by the ratio of daylight hours to night-time hours. In spring, the days begin to grow longer, while the nights shorten; the opposite occurs at the onset of autumn. During the summer, the days are long, and the nights short; ambient temperatures play little part in the ferret's determination of the season.

In captivity, we can control the photoperiod (quite literally the length of light time), thus fooling our ferrets into believing that it is a different time of year than it actually is. If you are breeding animals for a business, this is essential, and the principle is used with many species, including chickens, where we need a constant supply of eggs throughout the year.

Jills will reach puberty at between five to nine months from birth and in the first 'breeding season' after their birth. Hobs reach sexual maturity a couple of months before this breeding season. As spring approaches, the female will begin to come into oestrus (season). The physical signs are self-evident, but I will explain them.

The vulva, the female's sex opening, will swell, and protrude from the body by two to three centimetres; there will also be a discharge and the jill will have a much stronger smell. As in most female mammals, at such times the jill may become irrational and bad-tempered; it is not her fault, but caused by a hormone imbalance. She will stay in season until mated, or to the end of the summer, although the vulva's swelling will vary from week to week. The jill is not ready for mating until the vulva has swollen fully – usually about two weeks after the first swelling occurs. Attempts to have the jill mated before this time will result in injury to the jill, and no litter.

It should be obvious that all jills kept in oestrus for any length of time will not be as happy or as fit as they would be normally. An old country saying in the UK is that if a jill is not bred from every year, she will die at a very early age; there is a lot of truth in this, and every year many ferrets die needlessly. There is, however, no need for the jill to have a litter every year; this is explained in more detail below.

The hobs, too, change when the breeding season arrives. The hob's testicles enlarge and descend into the scrotum in the early spring and summer, allowing them to breed; for the rest of the year, the testicles are out of sight, in the hob's abdomen. It is a pointless exercise trying to breed from a hob whose testicles have not descended as, even if he can perform physically, his sperm will be infertile.

When you have carefully selected the hob and jill which you wish to breed from, you can either place them both in a neutral cage, or simply put the jill in the hob's cage. Within a very short time – usually no more than a minute or so – the hob will have hold of the jill's neck, and will be dragging her to a place of his choosing; this is usually, but not necessarily, the nest-box. Once there, he will enter her, and the couple will stay tied together for two to three hours.

The actual coitus – the physical act of mating – is extremely violent and aggressive, with the hob biting the jill's neck and dragging her around the cage, mounting her several times. The whole exercise is punctuated by screams of pain from the jill, but the whole process is essential to induce ovulation.

After this mating, both ferrets will probably spend time cleaning themselves, and may even eat and drink a little. If they are left together, the hob will repeat the whole process. I have found it best to leave the pair together for no more than twenty-four hours, since more time will inevitably lead to injuries of the jill's neck. This length of time is usually sufficient, but if you wish to be absolutely certain that the jill has been properly served, and will become pregnant, place the couple together again, about forty-eight hours later.

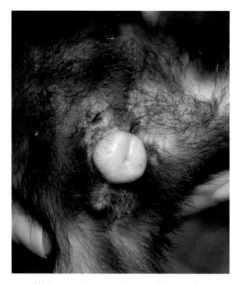

Jill in season – note swollen vulva.

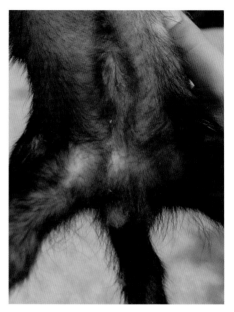

Hob in breeding condition – note descended testes.

After the mating is completed, inspect the jill's neck, and clean the wounds, dusting them with an antiseptic powder to help guard against infection. Within seven to ten days of mating, the vulva will begin to return to normal size, and this will be complete within two to three weeks. An experienced breeder or a veterinary surgeon will be able to palpate the tummy of the jill at about ten days after the vulva is totally reduced, i.e. returned to its normal size.

AVOIDING UNWANTED LITTERS

In the UK and some other countries, too many ferret kits are born every year; this causes many problems and often results in what can only be described as cruel practices – abandonment of kits, euthanasia and kits going to unsuitable homes. Although one could almost forgive our ancestors of a hundred years ago for such thoughtless actions, there is no excuse for a modern ferret keeper.

As stated earlier, our ancestors recognised that jills which were not bred from each year remained in season for up to six months, and rarely lived for more than three to four years, and often much less. This is a consequence of the high levels of oestrogen – the female sex hormone – which remain in the body of the jill while she is in season; the only way to stop the build-up is for the jill to be mated. This phenomenon is known as induced ovulation, and the actual ovulation occurs about thirty hours after the mating.

If not mated, the build-up of oestrogen may well lead to a disorder known as *oestrogen-induced anaemia,* which can lead to leukaemia and ultimately the death of the jill (see Chapter 9 for more details). It should be noted that it is the coitus and *not* the birth of a litter that is necessary to prevent the onset of this condition.

In 1981, I mentioned the problem of prolonged oestrus in ferrets to my friend, Chris Charlesworth MRCVS, and made the suggestion that a vasectomised hob should be able to induce ovulation in the jill without her becoming pregnant. After some discussion, Chris agreed to try it and he carried out what I believe to be the first ferret vasectomy. In his book *Fred J. Taylor's Guide to Ferreting* (Buchan & Enright, 1988, page 115), Fred mentions that he had heard about vasectomies in ferrets, and I believe that this could only have been the ones that Chris operated on. The operation was tricky but a complete success and Roger, as the ferret was aptly nicknamed, began a long career of taking jills out of season. The term for a vasectomised hob, hoblet, was one I coined in the first edition of this book, and is now in common usage around the world.

If you never intend breeding from your jill, the easiest way to avoid this problem is to have her neutered. However, if you may wish to have a litter in the future, as discussed in Chapter 10, it is possible to treat the jill with drugs. To my mind, the easiest and best method of avoiding unwanted litters and ensuring the health of your jills is to own a hoblet – a vasectomised male.

A hoblet has merely had the tubes which transport the sperm from his testicles to his penis cut, and so can perform exactly the same mating as a full hob. A hobble – a castrated male – has had his testicles fully removed, and so will not be capable of performing coitus at all. There is a world of difference between the two operations, but both are easily within the skills of a competent veterinary surgeon.

Some ferret keepers share the services of a hoblet, but this practice is full of potential dangers, not least of all the spreading of infectious diseases among ferret kennels. If you are serious about ferrets, then you should own a hoblet.

After having been mated with a hoblet, the jill may well go into a pseudo- (or phantom) pregnancy; this will last for the same length of time as a normal pregnancy (about forty-two days). At the end of this time, she may well come back into oestrus, and will require mating with a hoblet again.

THE BIRTH OF THE LITTER

The ferret's gestation period (pregnancy) is between forty and forty-four days, with the average being forty-two days. Where jills are kept in a commune, or several are kept together, it is best to separate a pregnant jill and place her in her own cage about seven to ten days before the litter is due. This will give her time to make the cage a home, and feel comfortable and secure in it.

Many people who keep several ferrets together, leave the whole business together all year round; with some, this causes no real problems. However, where there are several ferrets in the same cage, there is always a potential for trouble among the inmates, especially during the spring and summer months.

When spring approaches, hobs and hoblets will 'come into season', and become more aggressive towards each other, with the jills, and occasionally with the handler. It is not unusual for two hobs caged together to injure each other, sometimes quite seriously. For this reason, although I have often kept the whole business together, I now invariably remove all hobs and hoblets from the courts at the onset of spring, and keep each of them in their own cage. With the jills, I treat each one as an individual; I know that some are extremely tolerant mothers, while others will not accept another jill in the same cage at breeding time. It is important that there are enough suitable cages for all the jills and hobs during the breeding season.

Once the kits are eating solids and are self-sufficient, I often mix whole families, until such time as I have chosen which animals to keep, and which to let go. Even then, there may still be individuals who will not be able to tolerate each other; these must obviously be kept separately.

The signs of impending birth are obvious – nesting begins about ten days prior to the birth, and the mammary glands swell, with the nipples becoming obvious.

Early Days

Kits are born blind, deaf, naked and totally dependent on their mother, who feeds them with her own milk for the first six weeks of their lives, although the kits will begin eating solids from about three weeks – even before the eyes are open.

At birth, the kits weigh about eight grams, and it is important not to disturb the nest unnecessarily for at least five to seven days, although some jills, particularly the more mature ones with breeding experience, may not object or act at all adversely. When a jill feels that her litter is threatened, she acts in a way which may, to us, seem illogical; from her point of view it is perfectly logical.

When faced with the threat that a predator will kill and eat her young, the jill, who has put a lot of effort into the production of the litter, will feel like you or I would if

A litter of kits just a few days old.

asked to surrender our salary after having worked hard for it. She does not surrender her litter, but kills them and eats them; this turns the kits back into protein and also brings the jill back into season. This recycling of the kits helps wild polecats to ensure the continuation of the species.

Although the jill has only eight nipples, she can nurse more kits *provided that* she has sufficient food of the right type. All growing tissue requires protein, and being a parent calls for much energy, and so the jill should be provided with a high-protein diet, with sufficient fats and carbohydrates for her needs. If the jill loses a lot of weight, it may be because of an inadequate diet, although it is quite normal for a jill to lose some weight while nursing a litter of kits.

FOSTERING OF KITS

Very rarely, a jill cannot raise her litter; this may be because of illness or injury, or some other factor. Whatever the cause, you need to decide what you are to do with the litter. The ferret's milk is unlike cow's milk, and it is highly unlikely that you will be able raise the kits on this milk. A ferret's milk consists of 23.5% solids, of which 25.5% is protein, 34% fat, and 16.2% carbohydrate, and to rear young kits successfully, you will need to emulate this milk.

Some breeders have had a reasonable amount of luck with dog and cat milk substitutes, while I have raised ferrets from two weeks on a mixture of evaporated milk and water (1:2), mixed with the yolk of an egg. I feel that below two weeks of age, or where you cannot guarantee being able to give the one hundred per cent commitment to the task of hand-rearing the kits, this should not be attempted, and the kits should be humanely destroyed unless there is a viable alternative.

To help overcome such problems, and especially when the litter is extremely important, the use of a suitable foster-mother is to be recommended. I always try to have two jills giving birth on about the same day; in this way, if there is a problem

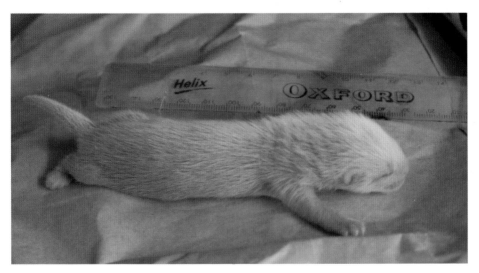

Kit eleven days with eyes and ears still sealed.

Kits in nest – eleven days old.

Kit fourteen days old. Handling of young kits should only be carried out if the jill is not worried by this action, and is not recommended for novices.

Kit twenty-five days old – furred but eyes and ears still closed.

A well-furred kit, aged thirty-two days, with its eyes just opening.

with one of the jills being unable to feed part or all of her litter, I can use the other as a foster-mum. Before the kits are placed with the foster parent, she should be distracted by placing some food in her cage, well away from the nest.

To ensure that she does not simply pick up the food and carry it back to the nest immediately, give a dish of sloppy (or even liquid) food.

As she is feeding, carefully and quietly place the kits in their foster nest, and sprinkle some shavings from the cage on to them; this helps mask their smell, making them more acceptable to the foster-mother. Hopefully, the mother will return to her nest and settle down to nursing without ever realising that her litter has grown. Do not place too many foster kits with her, as this may make her incapable of looking after any of the kits properly.

A jill in nest with her litter about thirty-five days old. Kits' ears open at about thirty-two days, and eyes slightly later.

WEANING

The kits develop quickly, as detailed below.

Fur erupts	9 days, profuse covering by 5 weeks
Eyes and ears open	3–5 weeks
Venture from nest	3–4 weeks
Weaning	6–8 weeks (about 300g–400g)
Canine teeth erupt	7 weeks
Adult weight reached	4 months (no longer classed as kits)

Many people try to give complicated methods of weaning the kits, stating that it is essential to provide finely chopped food or even liquidised food; this is totally unnecessary, and unnatural. In the wild, the jill simply catches her normal food and carries it to the nest, where the kits help themselves; you should emulate this behaviour. The only considerations necessary are the amounts, a high protein, fat and carbohydrate content, and the removal of all uneaten food before it has a chance to go bad.

Chapter 7

Working

As discussed in previous chapters, the ferret was probably first used by man to aid the hunting of rabbits or similar animals; today, although illegal in some countries, the practice still exists, and in the UK it is probably one of the main reasons for people to keep ferrets. It is not a sport for everyone, but it does have a lot of good points.

Rabbits are an agricultural pest which need to be controlled; many countries have laws which place responsibility for the control of rabbits and other agricultural pests on the owners and/or tenants of land, and not to control these animals is an offence against the law. When rabbits are caught by ferreting (ferrets do not fetch out the dead rabbits, and should not kill them at all), they are inevitably killed humanely and are then used as a food source, either for humans, animals or both. To my mind, this makes the practice incredibly 'green', since not only are we controlling a pest species, we are also providing good healthy meat. While alive, the rabbits live a perfectly natural life, and are not subjected to many of the excesses that domestic food species may suffer in captivity.

The rat is another species which has been successfully hunted and controlled by using ferrets, although the cadavers of such sport cannot be eaten, owing to the risk of disease. Ferrets used for ratting need to be smaller than those used for rabbiting, and should have gained at least two years' experience against rabbits before being used on the rat – a far fiercer adversary. One of the biggest dangers facing those who hunt rats is Weil's disease, or leptospirosis (sometimes called leptospiral jaundice), com-

A good bag: the author and a couple of his ferrets after clearing a rabbit warren.

The author between buries while conducting rabbit management on a farm.

monly known as the 'rat catcher's yellows'. This is caused by spirochaetes, a type of bacteria and, in common with other bacteria, once this enters your bloodstream (perhaps through a cut or graze on the hand), it causes a fever and produces toxic substances, often with fatal results.

These bacteria do not affect the host animal, the rat, which is therefore known as a passive carrier, and it is thought that the rat and the spirochaetes may have evolved together, since the bacteria require a host animal in which they can survive and which they do not damage, in order that the bacteria themselves are not wiped out. Some of the bacteria are passed out in the rat's urine and all that is needed is for a human to get some of that urine on a cut or graze, to be infected. Likewise, eating food with hands soiled with rat urine is a sure way of infecting one's body with these bacteria.

The first symptoms – fever, diarrhoea, severe thigh pains and vomiting – appear within a few days of the infection, and within a week the victim is jaundiced, owing to liver damage. It is usually at this stage that the family doctor is given some indication of the nature of the illness and, in order to treat the patient successfully, the doctor must move quickly; if not, and the patient's own immune system is not too strong, the patient may be dead within the next seven to ten days.

This information is not intended to frighten the reader, merely to inform of the possible dangers of hunting the rat. These dangers can be minimised by wearing strong rubber gloves when hunting rats and by adhering to a strict hygiene routine, i.e. always washing hands well before eating or drinking. Ferrets used for ratting will inevitably get infected urine on them at some time, and by handling these ferrets, you may well infect yourself. Ferrets cannot catch Weil's disease.

OBTAINING HUNTING PERMISSION

Whichever quarry you decide to hunt, you must obtain permission from the person who is legally entitled to grant it to you; this varies from country to country, and so the reader should check with a legal expert if in any doubt about who can grant such permission.

I have found the best way to get hunting permission is to keep a lookout for areas where there is a definite problem with, say, rabbits. Next, ascertain who owns or leases the land, and make contact by telephone to make mutually convenient arrangements to visit. When you go, do not wear your ferreting clothes, but go along in a respectable outfit; first impressions count.

Explain to the landowner that you believe he or she has a rabbit problem and that you would like permission to reduce the rabbit population with your ferrets. In return, you will supply the landowner with some of the dead rabbits, and will respect their property. Don't be too disappointed if you have several unsuccessful meetings before you get your big break. However, once you are given the chance, do the job well, be courteous and always offer the landowner a couple of rabbits from each trip. News of a good and efficient rabbit controller travels fast, and you will probably get more offers of ferreting. This is how I started, and today I have more ferreting than I can handle on a regular basis.

STARTING OUT

The amount of equipment needed for hunting rabbits with ferrets is minimal – ferrets and nets are sufficient, although, as we shall see, some other items of equipment will help us be more efficient and effective.

The ferrets chosen for hunting should be of average size; too big and they cannot manoeuvre enough; too small and they will not possess the reserves of energy needed for a full day's working. The colour of working ferrets causes arguments and discussion among many ferreters, with strong opinions being expressed in all directions by the parties concerned. The adherents of white ferrets will tell you that it is easier to see the ferret when it leaves the burrow although, as much ferreting takes place in the winter months, when snow is on the ground, this argument is not too sound, On the other hand, the adherents of the poley will tell you that these darker-coloured animals are keener to hunt and much more ferocious, making them the better colour for hunting purposes – an equally fallacious argument.

Similar arguments exist where ferrets are used in conjunction with hawks. Some *austringers* (those who hunt with hawks) will tell you that white ferrets must never be used, as they resemble the colour of the day-old chicks fed to the hawks, while others will tell you that *only* white ferrets should be used, so that the hawk will not think the ferret is a rabbit.

None of these arguments are true. My own theory for these tales is that, when the storyteller obtained their first ferret, they had no choice of colour, and now seek ways to justify their 'choice'. I have never noticed any real difference in the working abilities of ferrets which can be linked to the colour of their fur; if a ferret works well, it is the right one for the job, regardless of its colour and markings.

Another argument which rages with rabbiting ferreters is that regarding which sex is best. Many hunters will not use (or even keep) hobs, saying that they are useless for work; these people hunt only jills. Some say that most jills are too small to be able to work for long enough periods, and so only use hobs. It is true to say that the great majority of ferreters use mainly jills, with hobs only being used for specialised roles, such as a 'liner' (see later for more details). I must state here that one of the very best working ferrets that I have ever possessed was a hob which was rescued from a man who no longer wanted his ferrets; the hob worked long and hard, would not go underground if there were no rabbits in the burrow, and would not come out until the job was done. I was very sad when he was stolen from me (along with ten other ferrets) in February 1994.

The basic concept in ferreting rabbits is to find a burrow occupied by rabbits, and then introduce one or more ferrets into that burrow. It is best to work with at least two ferrets, as rabbits in a large burrow can easily give a single ferret the run-around. We hope that the ferret will not actually catch any rabbits, merely flush them out for us to deal with (although I'm sure that the ferrets have other ideas). The ferret is aided in this endeavour by its smell, and old-timers used to attach bells to the ferret's collar to scare the rabbits, and also to aid in finding the ferret.

TYPES OF RABBITING

There are four main methods of catching rabbits flushed from their holes by ferrets – nets, guns, hawks and dogs; by far the most commonly practised method employs nets.

There are several types of net used in modern rabbiting operations, and these are discussed in more details on pages 102–112. In general, different types of purse-nets and long-nets (in this context referred to as 'stop-nets') are used. Purse-nets are placed over every rabbit hole where it is possible to do so, to catch the rabbits as they flee the burrow, while stop-nets are placed around the whole burrow, to catch any rabbits that have managed to escape from the purse-nets. The two types of net should complement each other, and are best employed together, as it is all too easy to miss

A well-trained gundog (here an English springer spaniel) is a great asset and ensures that all rabbits shot are retrieved to the Gun.

A properly trained dog is indispensable to the ferreter. This spaniel has clearly marked a used rabbit hole.

an opening to the burrow, and the fleeing rabbit can then be caught up in the stop-net. Even where a purse-net is correctly set over a hole, it is not impossible for a rabbit to escape from it and, without the back-up provided by the stop-net, that rabbit would escape completely. In many cases, ferreters often simply use the stop-nets, and this can be a quick and easy way to catch conies, although I prefer to use both nets together, to ensure best results from my rabbit management efforts. (See also the section on nets.)

Where sport, and not the actual number of rabbits caught, is the main priority,

A female Harris' hawk is easily capable of taking on a fully grown rabbit. Note the telemetry transmitter on the hawk's foot.

Top-quality telemetry must be used at all times when flying hawks.

ferrets are simply entered into the burrow and the fleeing rabbits shot by carefully placed Guns. This method will not result in the numbers caught by using nets, as the noise of the shots will dissuade many, though by no means all, rabbits from bolting. It is also essential that all Guns (i.e. those people actually shooting at the rabbits) are one hundred per cent safe, and no shots are taken which may put at risk any human, ferret, dog or indeed any animal other than the quarry – the rabbits.

Using hawks to catch rabbits will give even smaller bags, but excellent sport. I do not intend to give the full details of hawking here, as it is a sport and subject in itself. The main principles in this sport are to have a suitably trained species of hawk, such as the Harris' hawk *(Parabuteo unicinctus),* which is held on the fist until the ferret flushes a rabbit from the burrow. The hawk is then released and catches the rabbit (see my book *Practical Falconry* for more details).

Many varieties of dog are used for hunting rabbits in conjunction with ferrets, and all have their adherents. Again, I can only skim over the subject in this book, as our main concern is the ferret, so I recommend that interested readers consult practitioners and books for further information.

In essence, there are two main types of dog used for catching the rabbits flushed by our ferrets – gaze hounds and terriers. Gaze hounds, or lurchers, hunt their quarry by sight and need open space to be successful; they work well where one is ferreting a bank on the edge of open fields. Terriers do not have such a turn of speed, but are much more manoeuvrable than lurchers, and can work well in undergrowth where lurchers could not operate. Both types of dog, however, have a use which any ferreter will appreciate; they can be trained to mark burrows which contain rabbits, thus eliminating time wasted by netting up and entering ferrets into empty burrows.

In addition, many ferreters, myself included, like to use dogs to find and indicate ('mark') warrens where rabbits are living. I use English springer spaniels for this, as these dogs have a superb nose and, if properly selected and trained, are a superb asset to the ferreter who wishes to maximise his/her bag efficiently. Our dogs are trained to use their nose to scent the rabbits and, if they detect the scent in a burrow, the dogs' tails (always active in springers) become even more animated, and this flag waving tells us that we will not be wasting our time by setting nets over a particular working.

TRAINING

A lot has been written about the need to train a ferret to do its job correctly, i.e. to hunt rabbits. Some books devote huge sections to this aspect of ferreting. To my mind, there is no need to teach an animal to carry out a perfectly natural function. The ferret is, by nature, a predator – a hunting animal – and, left to its own devices, it will do the job naturally. Undoubtedly, the best way for a young ferret to learn its trade is to follow the example of an experienced ferret but, if you do not possess such an animal, do not despair.

A rabbit's bladder should be emptied before the animal is skinned and gutted.

Most animals have an instinctive fear of the unknown and, for this reason, some young ferrets may be reluctant to enter a dark tunnel which heads down almost vertically. Choose instead, a reasonable size of tunnel in a burrow that you know is uninhabited and consists of no more than two or three holes. Ensure that the tunnel, initially at least, slopes slightly upwards and place the ferret just in the mouth of the tunnel. Do not worry if it pauses and sniffs the air for a time before venturing into the abyss, this is perfectly natural. Do not try to force the ferret underground before it is ready; to do so may cause the ferret to fear the underground world, thus rendering it virtually useless for hunting. Patience will pay dividends.

After one or two outings of this nature, take the tyro to a small burrow where you would expect to find a rabbit or two and enter it. Once a ferret has had a first encounter with the cast of *Watership Down*, there will be no stopping it. Remember that it will be very excited when coming out of the hole, and so do not try to pick it up before it has completely left the hole. To do so will encourage it to 'skulk', i.e.

Freshly killed rabbits hanging in a tree to cool. It is essential to allow rabbits to cool down naturally, in order to preserve the quality of the meat.

always wait just inside the entrance to a tunnel and move away from your hand every time that you try to pick it up. It is easier to start this habit than it is to stop it.

By the time your ferret has been entered to occupied rabbit holes a few times, it will be 'trained' and, in future, whenever you take the ferret out of its cage to put it into its carrying box, it will realise that it is probably going hunting, and may even tremble with excitement. This excitement will always be evident when the ferret is taken out of the carrying box on site, and introduced to the opening in the ground from which exudes the unmistakable aroma of rabbit. Your ferrets will look forward to such outings and consider them in exactly the same way that we do – not work but good, clean fun with (hopefully) a good meal at the end of the day.

With experience, ferrets will learn which burrows are occupied; I have had many ferrets which simply refuse to go into some burrows (presumably because there are no rabbits in them), and yet will dash into others with much gusto and enthusiasm. Rabbits are inevitably caught from burrows where the ferrets act in the latter manner.

EQUIPMENT

Ferrets

The number of ferrets required will be determined by the area to be worked and, of course, by the number you have available. As stated earlier, I will never attempt any bury with just a single ferret, and I always prefer to use a team of three – one jill and two hobs. Hobs and jills work in different ways and the two sexes should complement each other. Jills tend to be very fast and get on with their work, while hobs have a tendency to drag their feet and walk rather than run through the warren. The fast actions of the jills usually result in many rabbits being flushed from the warren within a few minutes of the jill being entered; however, she can be *too* fast and miss the odd rabbit lurking around a corner in the warren. The hob's slower demeanour helps him to detect (probably by nose) any such loitering conies, and he will deal with these accordingly.

In large warrens, I like to take along at least two teams; in a very large warren, the two teams are often worked together, but I usually use one team and then another, alternating them throughout the day.

Many modern ferreters have never used a traditional liner (as described earlier), and I believe this to be to the detriment of their sport. A liner is simply a hob ferret which is kept separate from other ferrets, and so does not get on with them. He is used when a free-working ferret 'kills down' and then 'lays up', i.e. refuses to leave the dead rabbit, or when a ferret is trapped, either by a collapsed roof or some other obstacle. His job is to seek out the kill, and then chase away the errant ferret, while he stays next to the dead rabbit. The old-fashioned method was to attach a long line to his collar, via a swivel. The other end of the line is held by the ferreter and, as the line is marked every metre or so, the ferreter will be able to calculate how far into the burrow the liner is. Since the introduction of electronic detectors, it is not necessary to use a physical line, but rather an electronic transmitter, carried on the ferret's collar (i.e. telemetry). The usage is exactly the same, but without the line to get wrapped around every root, stone and other obstacle in the tunnels of the warren.

I believe that it is foolhardy to work any ferret without telemetry, and I will cover the proper uses of this essential equipment later in this chapter.

Carrying Boxes

All ferrets used for the hunting trips need to be transported to and from the working area, even where a vehicle is used to get the ferreter and his/her equipment close to the workings. As the ferrets will be working hard during the hunting foray, they deserve to travel in reasonable comfort. When I first started in ferreting, my mentor, Graham Moxon, a first-class gamekeeper, told me never to carry ferrets in a bag or sack. He explained that ferrets thus carried were subject to a fairly uncomfortable

A wooden ammunition box converted for carrying ferrets.

time, with little protection from the elements. As rabbiting is traditionally carried out in the winter months (the old adage was that one could only hunt rabbits in a month containing an 'r'), the ground would inevitably be wet. By placing a sack on the wet ground, the sack (and thus its occupants) would be wet. This meant that, when the ferret was finally removed from the sack and placed in a rabbit warren, the cold, damp ferret was tempted to rest in the warm, dry surroundings of the working rather than chase rabbits.

There is also a great risk that ferrets in sacks will be accidentally kicked or trodden on by the ferreters going about their business, and this could easily lead to the injury or even death of the ferrets.

Graham favoured carrying ferrets in a converted wooden ammunition box, in which he made several holes for the ventilation, and I also acquired one of these (see photo).

The teachings of my 'keeper have stayed with me all my life, and I am absolutely against the use of ferret bags of any description – be they specifically made for ferret transport or merely empty potatoes sacks. I want my ferrets to work their best in order that I can perform at my best in rabbit management, and so endeavour to give the ferrets a comfortable mode of transport throughout their working day.

After having tried many designs of ferret box, some bought from country shows and game fairs; others having been designed and made by myself, I have experienced the highs and lows of ferret carrying boxes. Some of the ones I purchased were only made from extremely thin timber and simply stapled together; after only a little use, the boxes disintegrated. Others were much too small, and others far too heavy.

In the 1980s, I made a ferret carrying box based on an old East Anglian design – a shaped box which fitted around my waist and had three compartments – one in each side and one at the rear. The idea behind this shaped box was to enable me to carry it comfortably over long distances, while still being able to get to the ferrets when I needed them. In theory, and while standing still, this worked but, once used on a few rabbiting expeditions, I soon discovered that the design had serious flaws. The shape, although comfortable when standing still, was extremely uncomfortable when the box was being carried over long distances. I find that when I carry a box over one shoulder, the tendency is to keep putting pressure on one side of the carry strap; this leads to the box being carried at an angle, causing the edge of the box to jar in one's side – very uncomfortable over long distances. Some ferreters have recently 'rediscovered' this design, and many so-called 'bow-backed' ferret carrying boxes are now being produced. I have tried several of these and all have the same problems as did my original almost thirty years ago. When one considers that the design is old but has died out, it is obvious that the design is flawed – even though some modern day 'experts' believe differently.

After many years of using ferret carrying boxes of all shapes and designs, I believe that I have now developed a great design for a modern ferret carrying box. The boxes I make, sell and use, measure 350mm long x 160mm high with a depth of 240mm, making them suitable for comfortably accommodating a working team of ferrets (one hob and two jills). The boxes are top-opening, with a sloping lid, and are equipped with a long, broad nylon carrying strap and a small carrying handle. This combination makes it easy to carry these boxes for long distances while still being able to lift them over hedgerows, etc.

Ferret carrying boxes must be large enough for the number of ferrets carried, with sufficient ventilation.

The boxes are built from top-quality marine (exterior) ply, and glued and screwed to ensure strength. There is a single compartment, covered by a single lid. I am often asked why I don't make double-compartmented boxes; the answer is, to my mind, simple – they aren't worth the effort. A double-compartmented box needs either two separate lids, or a single sliding lid which moves from covering one compartment to covering the other. In the first case, water inevitably seeps in through the gap between the doors. In the latter, there will always be a time when both compartments are open enough for the ferrets to get out. To me, it is far easier – and far more straightforward – to have a single compartment into which a team of ferrets (one hob and two jills) can be placed. Putting more than one compartment in a ferret carrying box simply complicates the matter and increases the weight and cost of the box. If a ferreter only requires one ferret out of the box, this can be simply lifted out, leaving the other(s) behind.

Ventilation is of great importance, even in the winter months, and to this end, my

Rear of rabbiting vehicle showing equipment. Note the large ferret transport box.

boxes have a large hole in the front of the compartment, which is covered by 25mm square, heavy gauge weld mesh. When it is warm, I place a very small amount of straw in the box over a sprinkling of shavings (the latter to absorb water and urine), while in really cold weather, I put in more straw.

As my ferrets are expected to put in a full day's work, they will require food and water during the day, and my boxes are equipped with a small spring on the front onto which I can easily attach a water bottle when we stop for a break. The ferrets are also given a small handful of dry food in their boxes before setting out in a morning, and this is replenished throughout the day. Once the ferrets have been given their food and water, I use the carrying box as a seat while I enjoy my lunch without the necessity of sitting on the damp ground.

When carrying out large-scale rabbit management operations, it is necessary to have a large team of ferrets available, and for these a large carrying box, resembling a portable cage, is used. Even when I am only taking six to eight ferrets with me (I sometimes have to take up to twenty), I usually employ this box, which is big enough to house them all in comfort, has fittings for a water bottle, and can be stuffed with straw to ensure warmth, while still having adequate ventilation. When I first started to use such a box in the 1970s, many ferreters scoffed at the idea but, today, many realise the benefits to both ferrets and ferreters of using such a box.

Electronic Detector

When I was first introduced to ferreting, I was taught how to use a traditional liner. This was a hob which wore a collar to which was attached a long length of string (later nylon cord was used). The string/cord was marked at regular intervals to help give an indication of how far into the warren the liner had travelled, and the hob used for such operations was – as far as other ferrets were concerned – totally anti-social, although it had, of necessity, to be totally capable of being handled. The hob chosen as a liner also had to be bloody-minded, staying with the dead rabbit until dug out by the ferreter.

When the first electronic detectors became available in the 1970s, many ferreters decried them as being useless compared to the infallible liners they used. However, rose-tinted spectacles apart, the problem with traditional liners was that the line could easily become tangled on underground roots and rocks etc., and the tunnels in rabbit warrens do not go in straight lines, often curving around underground. I have experienced many instances of having fed a liner into a warren, where it dragged in over 10m of cord, only to find that the ferret was, in fact, a mere one metre directly under my feet.

The only accurate way to find a liner on the end of its line is to dig along the line. Ferreters of yesteryear used various implements to help them in this. First,

The business end of the probe, showing the enlarged piece a few centimetres above the point. It is this bulge which produces an oversized hole, and allows the operator to know when breakthrough to the tunnel occurs.

The author with two old-fashioned rabbiting tools – a Norfolk long spade (left) and a proggling iron.

they would use a proggling iron (the word 'proggle' means to prod or poke); this item is also called a proggle, prodder, or probe, and is basically a very large version of the garden dibber favoured by green-fingered gardeners throughout the world. A ferreter's proggle is made of iron, T-shaped, and about 1.2m long. About 75mm above the point (which is blunt to prevent injury to the ferret) is a large bulge in the rod. The proggling stick is pushed into the ground above where it is believed a rabbit tunnel runs. It takes a lot of strength and determination to get the proggle into the ground but, when the stick breaks into a tunnel, there is consid-erably less resistance (owing to the bulge), and this marks the fact that there is a tunnel. I was taught to tie a brightly coloured piece of rag to the handle, to prevent it getting lost – which is all too easy to do in thick undergrowth.

In theory, the methodology is simple; when a ferret 'lies-up' i.e. stays with the dead rabbit and thus refuses to come out of the working, a liner is introduced. The liner should (in theory at any rate) go straight through the warren's tunnels to the errant ferret and chase it away. The theory is that the free-working ferret will now exit the workings, while the liner curls up and lies by the dead rabbit, waiting to be dug out by the ferreter. Once the liner has ceased moving, the line obviously stays still but, while it was moving, the ferreter would have made a note of how much line had been taken in by the liner by reference to the marked line. When the liner stopped, and the free-working ferret had been safely boxed, the ferreter would look into the tunnel into which the liner had been introduced and try to work out the direction the liner had taken. The ferreter would then try to match the underground tunnel from above, and push the proggling iron into the ground to find the tunnel. At this stage, there would be no guarantee that this was the correct tunnel, and so the ferreter would reach for his 'graft', probably a Norfolk long spade, a device which was favoured by many ferreters. A graft is aptly described as a long spade, since the overall length was about 2m (see photo). At one end of the device is the digging blade, fashioned in one of two styles – a medium-sized narrow blade or a larger, rounded 'spoon', the former for hard ground and the latter for sandy ground. The blade is rounded and set at a slight angle to the shaft; this allows the user to extract soil from a deep hole, without losing most of it on the way out. In order to help cut through roots, many ferreters would grind a fairly sharp edge on the blade (something I still do with all my grafts).

Once the ferreter had found a tunnel by using the proggling iron, a hole would be dug with the long spade, to check that the line was in that tunnel. On the opposite end from the blade is a rounded hook; once the line was located, the ferreter would turn the spade around, and push the hook into the tunnel to snag the line, which was then lifted out of the tunnel to allow the ferreter to gauge the direction the liner had travelled. Once again, the ferreter would use his proggling stick to find the tunnel and dig another hole and thus repeat this procedure until he found the liner. If the liner was too deep for the ferreter to reach with his arms, he would use the hook to snag the line, and twist the spade to fasten the line to the hook, and then lift out the liner ferret. Once the liner was safely ensconced in its box, the hook would be used to lift out the dead rabbit.

In the hands of an experienced ferreter, the long spade has another use – it can become a listening device. The blade of the spade is pushed about 30cm into the ground above one of the warren's tunnels and, by putting his ear against the shaft,

Ferreting spades – note all handles allow comfortable grip even when extreme force is required. The short-handled spade in the centre is an ex-MoD spade and is the author's tool of choice for digging under hedges and in confined spaces.

the ferreter can hear the noises of rabbit and ferret running around underground. By putting the spade into the ground in several different locations, a ferreter can triangulate the noises to find the (almost) exact location of the warring pug and drummer.

However, once the new electronic collars had become established, the old-fashioned liner soon became a thing of the past. I obtained my first 'bleeper' in the mid-1970s, and my first Deben Ferret Finder in 1978 (see photo).

The Bleeper was invented by Stephen Alexander Horchler, a keen British golfer, in order to help him and other golfers find balls lost during a game of golf. The transmitter was sealed inside a golf ball and the signal picked up on a hand-held receiver. However, the golfing authorities thought such an idea unsporting and not 'the done thing', and banned its use. The story would have ended there, had it not been for Horchler's appearance on a UK TV programme about the latest technology, *Tomorrow's World*. When the programme was transmitted one of the millions of viewers was a Norfolk gamekeeper, John Lawrence, and he saw the potential for using the technology for finding lost ferrets. Through the BBC, Lawrence contacted Horchler's company, Euronics, and arranged for samples to be sent to him.

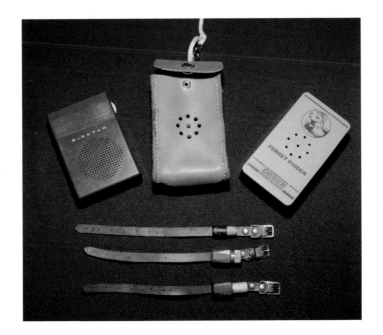

Early versions of electronic ferret detectors – the Bleeper (left) was based on an idea for finding lost golf balls.

Horchler dismantled the golf balls and removed the transmitters, one of which was then attached to a leather collar designed to fit a ferret, and the equipment was tested by Lawrence. Many problems were encountered during this testing, not least the fact that the original battery was designed to be a permanent fixture, with no means of recharging. Lawrence tried a rechargeable battery, but with no success, but finally adapted a co-axial aerial mount to take a small hearing aid battery, enabling the battery to be quickly and easily removed and/or changed.

When Horchler's company finally decided that there was no future in their golf ball finder, they happily sold all stock and intellectual rights to Lawrence, who kept the name Bleeper, but began marketing the gadget to British ferreters, to help them locate ferrets lost during rabbit management work. As a by-product, the device was also picked up and used by many workers in a wide range of industries – electricians and technicians who needed to thread cables; farmers and water-industry workers who needed to find and clear blocked drains, to mention just a few.

The Bleeper, good as it was in comparison to the old liner, was beset by reliability problems, and many ferreters who had such experiences refused to spend money on the device, and looked for ways to save a few pence. The Bleeper's receiver looked like a transistor radio, and many ferreters tried using these, which they tuned to the frequency of the transmitter and then glued the mechanism with epoxy resin to prevent it being detuned; this poor attempt to save money often led to a total failure to find the lost ferret.

Luckily, however, some ferreters pursued the idea of an electronic ferret finder with their technically minded friends, in order to try to get a reliable piece of equipment which would help them in their work. One such was Paul Walker, a telephone technician from Suffolk, England. Paul was, and still is, the driving force behind the Deben Ferret Finder, today considered an essential piece of equipment by all professionally minded ferreters.

First produced commercially in 1978, the Deben Ferret Finder went through various changes, many of them involving the internal workings, while the outer box remained almost identical, apart from the markings on the case. However, in 2004, a totally new ferret finder was launched by Deben – the Deben Ferret Finder 2 (FF2), and in the following year (2005), the original (now termed the Mark 1) was officially declared obsolete.

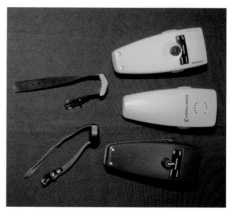

The Deben Ferret Finder 3 (FF3) is an essential piece of equipment for the modern ferreter. Note the pin pointer on the underside – the model on the bottom is a very early version, with the pin pointer being proud of the casing. In later models, such as the one at the top, the knob is recessed.

The new version featured a better (longer) range, a modern look, a brightly coloured case (although only on one side), a slim-line collar, waterproof magnetic reed switches (with no moving parts that could be affected by the adverse weather conditions, all too often encountered on rabbit management work), LEDs (light emitting diodes) combined with a speaker to indicate the depth/range of the collar from the receiver, a sophisticated anti-interference circuit to help minimise/eliminate the interference with which the Mark 1 had been blighted (it tended to give erroneous readings when used near metal fences and electrical cables, etc.), both a search and a locate function (see later) and, overall, better reliability. Unfortunately for all concerned, the FF2 had not been tested sufficiently, and suffered from many flaws, which obliterated any advantage that the new system was meant to deliver and caused a lot of bad feeling in the ferreting world, leading to many people building their own (unofficial) versions of the Mark 1.

The FF2 had extremely poor power consumption, leading to the early demise of the batteries in the collar, which simply died without any warning. In addition, the set was only capable of detecting one collar at a time and, as most ferreters use several ferrets in the same workings, and all should have collars fitted, this was seen as a fatal flaw and weakness, which inevitably led to the demise of the FF2, which was replaced by a much better version – the Deben Ferret Finder 3 (FF3).

As Deben had invested large amounts of time and money in the design and manufacture of the FF2, they decided to keep the same design, and merely label it as the FF3. Many ferreters did not believe that the new device was, in fact, new but thought it merely a newly branded version of the old FF2 – which it most definitely is not. However, this perception exists even today and sadly prevents many ferreters trying the FF3.

The Deben Ferret Finder 3 is, in fact, state of the art, and features a weatherproof

Albino jill
with FF3.

receiver case, with top quality LEDs and a speaker tone emitting noises it is very diffi-cult to ignore. The early models had a simple 'locate and search' facility, which on later models was upgraded to include a 'pin pointer' control, designed to allow the operator to adjust the receiver's sensitivity and mute the sound until the ferret wearing the collar was extremely close to the transmitter (in much the same way as on the Mark 1). As the ferreter closes in on the ferret collar, so the LEDs change, as does the pitch of the tones emitted by the receiver. By engaging the pin pointer, the set can be used with only the LEDs flashing, and no noise – an obvious advantage at times.

To address the problems with battery life in the FF2, the FF3's range was reduced slightly, from 6m to 4.9m; I expect about thirty hours of working from each battery used in the receiver, with the collar batteries giving about ten times this duration (note that the batteries in the collar are switched on at all times when the collar is worn by the ferret, whereas the receiver is only switched on when searching for the ferrets). The circuitry in the receiver has been changed and improved from that utilised in the FF2, to allow several collars to be used at any one time with a single receiver. The FF3 retains interference-reducing circuitry, which effectively filters out many erroneous signals which would have been picked up by the Mark 1, and which could lead to a loss of valuable time by the errors thus produced.

Deben continue to make modifications to the FF3, with the hope that it will improve the performance of the unit, thus giving ferreters a device that will aid their work. The latest such development is the FF3M; the M denotes a magnetic switch which is used to turn the collar on and off. The FF3M collar has a screw lid and is designed to be waterproof; unlike the original FF3, which has a nylon collar, the collar of the FF3M is leather. The design allows the collar batteries to be left in situ, being turned on or off by the use of a magnet which is supplied in the set or is incor-porated in the latest receivers (see photo).

My main criticism of the new FF3M is the size of the transmitter. While some ferreters use only hobs for rabbit management work, we use both hobs and jills. As the jill is so much smaller than the hob, this makes the collar too big, heavy and bulky for many of our smaller jills. Nor do I like the idea of trusting a magnetically operated switching mechanism on the collar. This can all too easily be switched off accidentally during use, and there is no obvious way of knowing if this has happened when working a ferret wearing such a collar.

The lack of waterproofing on the original collars of the FF3 can be almost completely overcome by the use of insulating tape wrapped around the collar once the batteries have been fitted and the collars checked for correct operation. The small, slick size and shape of the FF3 collars are, to my mind, far superior to the bulky and heavy FF3M collars. It should be noted, however, that at the time of writing (December 2011) Deben no longer manufacture the FF3 collars, only the FF3M collars.

Guide to the Deben Ferret Finder 3

While some ferreters steadfastly refuse to use the new technology, I believe that only a fool would work a ferret without using an electronic ferret finder and, although I understand that no system is foolproof or 100% effective and efficient, I believe that many ferreters are put off using the FF3 as a result of problems with the FF2 (and the erroneous perception that the FF3 is the same technology – it is not), but mainly by the lack of understanding as to how to use the FF3 effectively. With this is mind, the following should prove useful to all users of the FF3 and FF3M.

PREPARATION

Remember to use the correct batteries, which need to be in good condition. The FF3 and FF3M receivers use a 9-volt PP3 battery; Deben recommends that the FF3 collars use two 393 silver oxide hearing aid batteries (other equivalents include R393/15, RW48, SR48, SR48W, V393), and the FF3M uses two 394 (silver oxide) types. Please be aware that the voltage of the batteries used has a major influence on the performance of the equipment. The Deben equipment calls for batteries of 1.55 volts, but it is possible to purchase batteries (usually much cheaper than those

A digital multi-meter is useful for checking the voltage of batteries.

recommended and supplied by Deben) which look identical to the recommended types but are, in fact, of lower voltage. Zinc air batteries should never be used – stick to the recommended (silver oxide) batteries, and avoid cheaper alternatives, as these will inevitably lead to poor performance, and even failure, of the equipment. Without the proper batteries, in good condition, the equipment cannot and will not function correctly.

When fitting the batteries, check for clean contacts (they should be shiny, with no oil, grease or rust present) and ensure the insides of the case and collar are dry. If there appears to be moisture present, wipe with a tissue or dry cloth and spray lightly with a water displacement spray (e.g. WD40). Check that batteries are inserted correctly and make good contact, and then switch on the receiver (and collar/transmitter if using an FF3M collar). The batteries must be fitted observing the correct polarity (i.e. positive + to positive + and negative – to negative –). With the collar batteries, the more pointed end is the negative, and the larger end is marked with the + sign; PP3 batteries are also marked with the + and – signs on the appropriate terminals.

Once a battery has been removed from its manufacturer's packaging, I store it in a small plastic box with separate compartments, designed to be used to store pills, and available from chemists and drug stores, etc. Keeping the batteries in separate compartments will prevent the accidental shorting of the batteries, thus preserving their lifespan. Before fitting such batteries that have already been used, I first check the voltage with a digital multi-meter (see photo).

Check the operation of each collar at varying distances, starting by placing one collar approximately 150mm directly beneath the receiver's pointed end (i.e. the opposite end from the LEDs). At this distance, the 0.5 foot LED should be flashing continuously, but no others. If the low LED does not flash continuously, this indicates a low battery, which should be changed to ensure optimum performance of the equipment. If the collar does not function at all, check the polarity of the batteries that have been fitted and correct if necessary. If the batteries are fitted correctly and are in good condition, but the collar still does not function, check that the terminals are clean and free from rust. I use a small piece of glass-paper to clean rusted terminals lightly; the rust is usually caused by the ingress and presence of water in the equipment. If the collar does not operate correctly despite all of the above checks, it should be returned to the manufacturer for repair and not used until given a clean bill of health by the technicians.

I recommend that all batteries should be removed from the equipment when it is not being used; this helps prevent damage caused by damp or leaky batteries, and is a rule I employ even with the FF3M. It is also good practice to check the operation of collars during rabbit management work, and always take plenty of spare batteries on rabbiting forays.

USING THE DEBEN FERRET FINDER 3 AND 3M (FF3 AND FF3M)
To use the FF3 to find a missing ferret, the receiver, having been properly equipped with batteries and tested as detailed earlier, should first of all be switched on and then the device switched to SEARCH; this will put the receiver on at full power, thus ensuring the greatest range. Once in the area where the ferret is thought to be, the

Using the FF3 to find a missing ferret.

Using the pin pointer on the FF3; this enables the ferreter to find the exact location of the missing ferret before digging.

pointy end of the receiver is pointed in the general direction of the lost ferret, and is waved slowly from side to side, keeping it pointing at the ground. If the ferret is within range, once the receiver is pointing in its general direction, tones (pips) will be heard from the receiver and LEDs will flash; the tone of these sounds will fluctuate with the range of the collar/ferret from the receiver, i.e. when closer they will be of a higher pitch, and they will emit a lower pitch as the distance between collar and receiver increases. Move in the direction of the higher tones and then swing the receiver in a left-right and up-down movement, noting where you get the highest pitch, and keep moving in that direction until the pitch changes very little when the receiver is moved around. Now change the receiver into LOCATE mode; this will make the receiver much less sensitive and therefore more accurate.

Note the ranges given over and below the LEDs – the top ranges are for use on SEARCH, while the ranges on the bottom of the scale are for use on LOCATE. All ranges are given in imperial measurements i.e. feet. For those not used to these quaintly old-fashioned measurements, the following conversions will be helpful:

IMPERIAL	METRIC
16 feet	4.9m
12 feet	3.7m
8 feet	2.4m
4 feet	1.2m
2 feet	0.6m
1 foot	30 cm
0.5 feet	15 cm

At this stage, the pointy end of the receiver should be directed straight at the ground and moved left-right and forwards-backwards until the pitch is at its highest. If using one of the later versions of the FF3, this is the time to engage the pin pointer, which will need to be set to zero by turning it completely anticlockwise (on older models of the FF3, a small click will be heard when the pointer is turned to zero) before any search or locate work is carried out.

The pin pointer merely reduces the sensitivity of the receiver to make it able to localise the readings of the collar, thus making it easier to dig for the ferret. As the pitch gets higher, turn the pin pointer until the noise just stops, and then move the receiver again until the noise starts again. Repeat this until the noise is only emitted when the receiver is held over a very small area of the warren, and take the reading off the receiver, allowing a few centimetres for the depth of the rabbit tunnel the ferret is occupying, and also the distance above the ground that the receiver was at the time of the reading. This is where your ferret is.

Avoiding the FF3 Tones
Some ferreters loath the noises made by the FF3, while others (myself included) know that there will always be times when the noises emitted by the FF3 are not needed, or can even be detrimental to rabbit management operations. To avoid the noises is simple, and is one of the reasons that Deben fitted the pin pointer; the pointer knob is turned completely anticlockwise. By doing this in both SEARCH and LOCATE modes, it is possible to prevent any sound coming from the receiver, which will then only indicate the presence of the collar/ferret by its LEDs.

Fitting the Collars
Unlike many animals such as dogs, cats and humans, ferrets have no obvious neck where a collar can be fitted, and it can be difficult to fit a ferret with a collar which will stay on amid all of the rigours of rabbit management operations. A correctly fitted collar will be a snug fit, whereby the ferreter can, with a little difficulty, move the collar around the ferret's neck but cannot pull it over the animal's head. If the human hand can remove the collar over the ferret's head without the collar being unfastened, then the collar is highly liable to come off during rabbiting work. It should be realised that all ferrets will differ slightly in size and build from each other, and that collars will almost definitely have to be altered to fit them. With leather collars this is simply achieved by using a proper leather punch to add more holes, but with a nylon collar, this is best achieved by using a red-hot nail of the correct diam-

A jill being placed under a hedgerow to bolt rabbits into a stop-net. Note ferret detector collar fitted – it is the height of stupidity to work a ferret without such a device.

eter; the nail should be mounted in a wooden handle to protect human fingers from the heat. This device will melt a hole in the nylon, and seal the edges to prevent the hole enlarging.

To fit the collar, I choose one which has been altered to fit the individual ferret and thread the end of the collar through the first part of the buckle fastening, thus making a loop. Holding the ferret in one hand and the looped collar in the other, I pull the collar over the animal's neck and tighten the collar to fit, testing to ensure the correct fit. When I am happy with the fit on the ferret, I thread the end of the collar through the other part of the buckle fastening and secure the loose end under the rubber band I have already fitted on the collar.

With FF3 collars, I always wrap the collar's receiver with insulating tape once the batteries are in; this helps prevent ingress of water and dirt, while also preventing the cap from being pulled off (which can happen if the receiver gets snagged and the casing broken). I also place a small rubber band on the collar (leather or nylon) which I use to secure the loose ends of the collar to prevent them snagging during operations.

A Few Tips

It is worth noting that for *any electronic ferret finder*, different soils will have a different effect on the readings given by the equipment; heavy, wet clay will give higher readings than dry, sandy soil for ferrets at the same depth. As one gets more familiar with using the ferret finder in different soils, so one gets the experience necessary to be able to make mental adjustments to the equipment readings to ensure accuracy in all soil types.

With the old Mark 1, in order to increase the depth that the receiver would pick up the collar/transmitter, it was necessary to hold the receiver as close to the ground as possible. This was easily achieved by attaching the receiver to a stick or a length of string, which enabled the unit to be held at ground level without breaking one's back. With the increased range of the FF3 and FF3M, this is not necessary, but it should be remembered that the closer to the ground the receiver is held, the greater

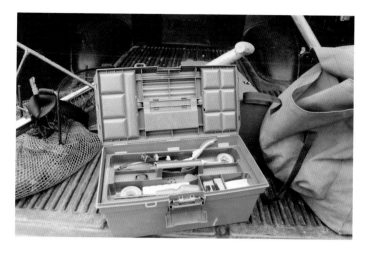

Using a single box to house all the electronic equipment, batteries and tools needed will be very useful to the ferreter.

the depth of soil it can penetrate, utilising the increased range of the FF3. When I have mentioned this to ferreters, I have often got the riposte that no one would want to dig five or more metres to rescue a ferret. I agree, but at least with the use of the ferret finder, one still knows where the ferret is, and that is always a great sense of relief to me when out working my ferrets.

Care of Equipment

When one has made a not inconsiderable investment in equipment, it pays to look after that kit, and I always store my ferret finders in padded cases designed for fishing reels. These cases are stored away from damp and excesses of heat, and will also house

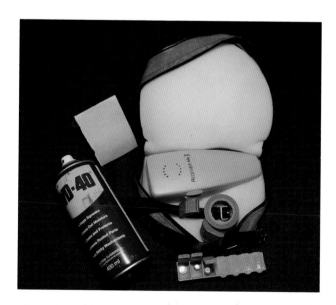

A ferret detector care kit – keeping the kit in a padded case (designed for fishing reels), will help protect this equipment.

Where digging needs to be done in an open field or ditch, a longer spade is useful.

the collars and boxes holding the spare batteries for both collars and receivers. I also use small tool boxes to hold in one place all of the ferret finder equipment – receivers, collars, spare batteries, hole punches, insulating tape, tools, spare collars, etc. Keeping all of one's ferreting paraphernalia in one place helps ensure that all items are easily to hand during ferreting work.

Recovering a Lost Ferret

Once the ferret is found with the receiver, rather than starting to dig immediately, it is worth trying a little trick to ensure that the ferret genuinely cannot exit the workings, or is simply grabbing some zzzs (sleeping). When I have located the ferret with the detector, I mark the spot (usually with a couple of sticks) and then quietly retire

Digging under a hedge requires the use of a short-handled spade, such as those previously issued to infantry soldiers in HM Armed Forces; note ferret detector placed to alert the ferreter to any movement of the ferret underground.

from the immediate area for about five minutes, before checking again with the receiver. If the ferret has not moved (or only moved a little) I stamp on the ground or hit the ground with the flat of my spade/graft. I check with the receiver to ascertain if the ferret has moved at this point and, if it has, it is obviously not in need of being dug out. If it is still in the same place, I know that I will have to dig, but I also know, from experience, that it may still decide to move as I dig down into the warren. To alert me to such movements, I place the receiver slightly to one side of the ferret's position, and set it to just pick up the signal from the collar. Now, if the ferret moves, the signal will alter and I will be able to avoid needless digging.

I use an ex-army spade for most of my rabbiting work. These come in two sizes – long- and short-handled; the long-handled were used with vehicles, while the short-handled spade was meant to be issued to infantry and carried on the personal kit/Bergen. I prefer the short-handled type as this allows me to work with it even in confined spaces (e.g. under hedges and buildings, etc.). In order to make digging easier, and help me cut through the inevitable roots, I sharpen the working edge of the blade, and paint the handle in a bright yellow or orange to help prevent its loss. I have tried the 'latest' spades with folding handles but find them prone to breaking on the hinges. I believe this is because the hinges harbour moisture and dirt, which will then corrode the hinges, leading to their failure.

Other Tactics for Recovering Lost Ferrets

Some ferreters prefer to use free-working ferrets without transmitters and, instead, utilise a liner, as detailed earlier. I prefer to use all of my ferrets with radio transmitter collars attached, thereby keeping me in touch with all of the subterranean happenings.

Some ferreters refuse to invest in ferret finders, stating that they have never lost a ferret – if you hear a ferreter making this claim, you are either talking to a ferreter who has done very little ferreting, or a liar.

In order to help retrieve ferrets without the use of a detector, other methods are sometimes used. The first is to condition your ferrets to a noise which indicates something good – usually food. This is the basic principle of all animal training and is best exemplified by experiments carried out by Ivan Petrovich Pavlov, a famous Russian psychologist. Pavlov conditioned his dogs so that, when they heard a bell ring, they could expect food. Rather than a bell, ferreters tend to use a set of keys, a call, whistle or squeak made by the mouth and, once a ferret fails to emerge from a warren, will make the requisite noise and expect the ferret to emerge. While this sometimes works (i.e. when the ferret is able to move out of the warren), it cannot work when the ferret has become trapped, whether by a roof fall underground, or by a rabbit. If the ferret has a rabbit pinned in a stop, it would be most unwise to persuade it to leave the coney, as it will learn that, by not doing its job, it will receive food treats.

A variation on the forgoing is to use a freshly caught rabbit. The dead rabbit's stomach is cut open and the smell is allowed to permeate the warren (often aided by the ferreter waving their hands around to help the smell circulate in the warren). Once again, this can only work if the ferret is able to move around the warren and is not trapped. It will also risk training the ferret to wait for this treat before exiting the warren, and doesn't work with a ferret fed purely on dry food.

A mink trap can assist in the recovery of missing ferrets.

If, despite all efforts, it is impossible to retrieve a lost ferret, I use a mink trap. This is a cage made from heavy duty wire and designed to catch – without harm – mink or similar shaped/sized animals (see photo). The mink trap is baited with food for the ferret, and then the trap is wrapped in hessian sacking and placed just inside the warren. The hessian sacking covers the trap's top, bottom and sides, but leaves both ends open (except for the trap's wire netting). Great care must be taken to ensure that the mechanism is not impeded by the sacking. The trap is designed to allow the ferret easy access to the food but, while moving into the trap to get the food, the animal must walk on a treadle which operates a locking mechanism thereby securely trapping the ferret. The sacking helps prevent the ferret from getting too cold (rabbiting operations being carried out in the winter months). The trap is usually only set when it has become too dark to continue the efforts to find the ferret and it must be checked first thing the next morning and at least every twelve hours to ensure the welfare of any animal thus trapped.

The flaw with the use of the live catch trap is that it can only work if the ferret is able to make its way out of the warren. If I have set a live catch trap and find it still empty when I return to the warren the next morning, I use the ferret finder to try to locate the errant ferret. I consider my ferrets to be the most important part of my rabbit management team and will go to great lengths to ensure their safety; I believe that using the latest ferret finding technology (currently the FF3) is the best way of doing this.

Whichever method you use, remember that practice makes perfect, and always ensure that you practice with your kit before you need it for a lost ferret.

Nets

As discussed briefly in an earlier section, there are several different types of net used by modern ferreters, and these can also be made from different materials. All have their adherents and critics, and I feel the selection of these nets to be very personal, while the choice can also have a huge influence on the success of one's ferreting activities.

Purse-nets

Arguably the most commonly and widely known of the nets used by ferreters to catch rabbits, the purse-net's name is self-explanatory – it works by pursing (closing) when a rabbit runs into the net.

The author setting a net over a bolt hole in a ditch. It is important to ensure that every hole is netted to prevent the loss of any rabbits.

Essentially, a purse-net is simply a square(ish) piece of meshed material with a cord around the outside, which forms the purse or bag, and is held in the ground with a peg, to which the cord is tied. Purse-nets usually have a stainless steel ring, about 25mm diameter, at each end, attached to the netting and through which the cord passes. However, the choice of material, mesh size, overall size, shape and quality of manufacture will also influence its effectiveness. Good-quality nets have individual

To set a net, it must be fully opened and then positioned to cover the hole.

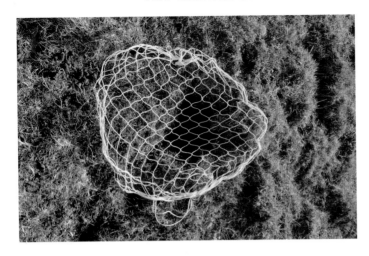

A correctly set purse-net – note high-visibility net and peg.

strands of netting securing them to a stainless steel ring at either end, whereas cheaper, inferior nets are tied with thin cord.

Poke-nets

A poke-net is an adaptation of the purse-net, and designed for covering multiple holes. The main differences between a normal purse-net and a poke-net are the size (poke-nets tend to be bigger – up to 1.5m) and the fact that the cord around the purse-net is not fastened to the netting at either end, but has a peg at each end of the cord.

Setting a purse–net: the peg is secured in the ground at an angle away from the pull.

Rabbit caught in a purse-net. Such rabbits can easily get out unless the ferreter takes swift action in securing and then dispatching them. Note high-visibility net and peg.

The design ensures that, no matter which angle a rabbit hits the net, the net will purse, running from either end, or even both. I highly recommend that every ferreter has several of these in the net bag. Once you have used poke-nets, you will find them indispensable.

A poke-net – note two pegs (one at each end) and high-visibility coloration of netting.

Long-nets

Long-nets can be very long – 100m is not unusual – but may also be much shorter, when they are usually referred to as stop-nets or gate-nets.

The nets are made of a length of netting with a cord or line along both long edges, i.e. top and bottom. Because this line will put a great deal of tension on the netting, the material on the outer edges is double thickness, and these lines of double thickness netting are known as selvage lines. In use, the netting is held upright by the use of stakes, traditionally made from hazel.

The nets are designed to stop running rabbits by getting the conies tangled in the netting and, to achieve this, there has to be excess netting (referred to as bagging); this usually equates to twice the length of netting compared to the length of line, i.e. a 100m long-net will cover 100m, but will be made up of 200m of netting. Without this extra bag, the net will not work.

The first long-nets I used were what is now termed 'traditional' and consisted of the net and separate hazel pegs. Setting was a laborious and fairly skilful endeavour, and one had to have a lot of experience to set and use a long-net properly. Tension was kept on the net by the use of an anchor pin at each end. If more than one net was needed, the ends of two were set to overlap to ensure there were no gaps through which rabbits could escape.

A long-net is traditionally used at night, when the rabbits are out of their burrow, and is set up (ensuring total stealth) between the grazing rabbits and their warren. Once in position, the rabbiters walk slowly towards the rabbits, making the conies run for the apparent safety of their warren, only to run into the long-net blocking their escape. A rabbiter will remain at the net, and his job is to dispatch the rabbits as they become entangled in the bag of the net, with the dead rabbits being gathered at the end of the event. Some rabbiters use dogs for flushing the rabbits to the net and, as a young boy, I helped a rabbiter flush the conies by using what was known as a 'dead dog'; this was a long length of rope, held at each end by one of us, and pulled slowly across the grass. As the rope moved over the grass, it emitted a weird sighing noise, and this caused the rabbits to move towards the net at a good speed without frightening them all into a mad gallop.

Stop-nets

A stop net is a long-net used in conjunction with daytime ferreting, and it can be used on its own, or in conjunction with purse-nets.

A stop-net comes into its own when the ferreter cannot see or physically get to every exit from the warren, e.g. when the holes are under heavy cover, bramble, hedgerows or buildings. The stop-net is placed around the areas from which the rabbits are expected to run, and from then on work in exactly the same way as the night-time long-net.

Quick-set Nets

The traditional methods of setting long-nets (and stop-nets) is laborious and difficult, and many ferreters have tried this and decided it was not for them. The invention of quick-set nets has helped more rabbiters to realise the potential of using long- and stop-nets, while making it extremely quick and easy for them to do so.

Correctly set stop-net having done its job. The author is removing dead rabbits from the net after the ferrets have been taken out of the bury.

(Left:) Setting a quick-set stop-net: the design of these nets enables the ferreter to deploy and set the net with great ease, even in adverse conditions.

A quick-set long-net in its carrier.

The quick-set nets have the pegs, usually made from fibreglass, permanently fixed to the netting, and the whole thing is carried in a very comfortable cradle. The use of this cradle enables the ferreter, while walking backwards, to set the net with very little effort. Since trying one of these nets about twenty years ago, I have not used the traditional type.

Gate-nets

The use of gate-nets is often overlooked by the present-day ferreter, and I believe that this adversely affects rabbiting operations. Gate-nets have many uses, their initial one being to cover gateways to prevent rabbits escaping through them. To set one, the gate must be fixed in the open position and the net set by attaching it to the gateposts on either side of the opening using the top and bottom lines. The gate-net functions like any long net.

These nets can also be used across ditches and openings in hedges.

Trammel-nets

A variation of the long-net, a trammel-net consists of three sheets of netting in layers. The outer layers are of small mesh (about 100mm) with the inner having a mesh of about 200mm. As a consequence of this design, when a rabbit (of almost any size) hits the net, it becomes entangled and held until the ferreter can get to it. I find this type of net extremely useful when working on my own, as I do not have to keep running up and down the length of the net to dispatch the rabbits, but can do so almost at my leisure, knowing that it is highly unlikely for any to escape once enmeshed in the netting.

The author's net bag, showing assortment of coloured nets. The bag must be well made, waterproof and big enough to carry a large number of nets.

Hemp and high-visibility purse-nets at a distance, clearly showing the advantages of using brightly coloured nets.

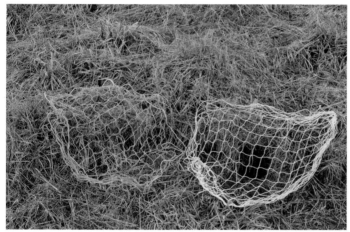

As rabbits are colour-blind the brightly coloured nets are just as effective as others, but much easier for the ferreter to see.

Colours

Many ferreters have fixed views on the 'best' colour nets for use in ferreting. Some opt for 'natural' colours (brown, green, etc.), arguing that these do not scare the rabbits, while others plump for bright, obvious colours (red, orange, etc.), contending that such nets are easily seen by the ferreter, and thus not left behind. As rabbits are colour-blind, the colour matters only to the ferreter, and I recommend that one uses the colour that one feels works best, as I am a great believer in the placebo effect of such choices.

Personally, as I replace my old hemp nets with those made from spun polyester, I am choosing very bright-coloured netting.

Mesh Size and Shape

While some ferreters will choose nets with a mesh size of as little as 50mm, others will elect to use mesh sizes as big as 150mm; I use mesh sizes of 55–75mm. When

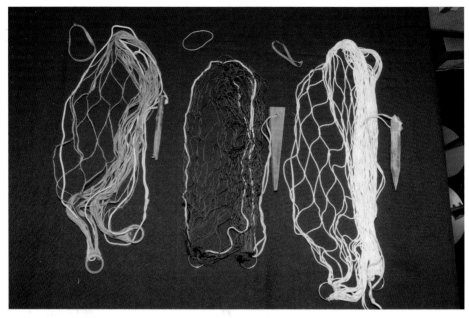

Purse-nets of different materials: from left, hemp, 6Z nylon, and 12Z nylon.

choosing the mesh size, take into consideration the material, as some (e.g. spun nylon) will stretch far more than others, such as spun polyester, with natural hemp having little or no capacity for stretching.

Materials

HEMP

Until fairly recently, hemp was considered to be the best (some would say only) material for quality nets. The material is strong, flexible, and easy to work (i.e. for making the nets), and easy to work with (i.e. it lays well when set over a rabbit hole). However, it also needs plenty of attention in order to make it last. After use, hemp nets must be cleaned (sometimes requiring washing in warm soapy water) and thoroughly dried before being stored. Many ferreters had their own 'secret recipe' for treating the material to ensure a long working life (usually some mixture of one or more wood preservatives). Hemp nets are also very expensive.

Many years ago I made my own nets from hemp and, some thirty years on, many are still in regular use. However, as any become damaged or need replacing for what-ever reason, I am, as mentioned, replacing them with nets of man-made materials.

NYLON

When nylon purse-nets first appeared, back in the days when hemp was king, nylon was seen as an inferior material, suitable only for youngsters and beginners, and those who simply could not, or would not, pay the extra money for hemp nets. This percep-tion was not helped by the fact that only very lightweight nylon was used for net making.

Nylon is measured in Zs, with one Z being the equivalent of 4kg breaking strain. The common weights used to be 4Z and 6Z, but very few used the 4Z nets, as the material is so light that it falls off rabbit holes and tangles in the slightest of breezes. Nets made from 6Z were seen to be adequate, but still considered second rate compared with hemp.

A few years ago, 10Z and 12Z nylon became widely available, and nets made from such material are excellent, giving long years of hard work, with few tangles. Nylon is, however, rather elastic, and so mesh sizes on nets (particularly long- and stop-nets) should be no more than 55mm.

SPUN NYLON

This synthetic twine is another recent innovation used for net making; however, it has an annoying tendency to twist and curl, and I find this makes nets made from such material very difficult and frustrating to use.

SPUN POLYESTER

This is another synthetic material which, although it has only been around for a short time, is becoming increasingly popular with ferreters who know their stuff. Being man-made, polyester doesn't soak up moisture like hemp, and is smoother to the touch, making it very pleasant to work with in both net making and in rabbit management operations. I have been using some of these nets for the last few years, and find them extremely hard-wearing, and this material is rapidly becoming my favourite for my nets.

Cords

All nets are attached to cord, which goes around the edges in purse-nets, and along the top and bottom of long- and stop-nets. The quality of this cord is often overlooked by ferreters, but if it is not of the correct type, weight and length, the nets will not function as well as they should. If a net's cords are too long, it is easy to trim the cord or, if only slightly too long, the excess can be twisted around the peg(s). However, if the cord is too long on a purse-net, it will lead to slow pursing, and thus give the coney a chance to escape.

Pegs

All nets are held in position by the use of pegs, and some thought should go into their choice. It should also be remembered that different types and materials will work better in some soil and weather conditions.

WOODEN

The traditional pegs were made from hazel, a tree which grows with multiple trunks which are very straight. Lengths of this wood were cut in the autumn and left for about six to nine months to dry out, at which time they were cut to the required length and one end was sharpened. Holes were drilled as required, through which cord or line was threaded. To many this seems very old-fashioned and, therefore, wrong. I do not consider this to be the case, and still make many of my pegs in this way.

NYLON AND FIBREGLASS

With the advent of cheap and plentiful man-made materials, many ferreters today use plastic, nylon and fibreglass pegs. I have tried these and will not use plastic or nylon, but find the fibreglass pegs very useful, particularly in hard ground. The difficulty with this material is fastening the cords and lines to the pegs, and I utilise castration bands (small rubber bands used to remove the testicles of farm animals) to secure the nets to the pegs.

Albino hob wearing a collar.

STEEL

Steel tent pegs and meat skewers are useful in cold weather, when the ground is frozen hard, and in areas where the ground is naturally hard, making the use of wooden pegs very difficult, and often impossible.

Other Items

Knives

Both a sheath and a lock-knife should be carried on hunting trips; the sheath knife used for hacking away foliage or cutting branches to use as probes, and the lock-knife for gutting the rabbits. Ensure that both knives are kept sharp, as a blunt blade is worse than useless.

Collars and Line

Collars for liners should be made of best-quality soft leather, kept supple by the use of a proprietary saddle-soap or leather treatment. The line should be of nylon, and at least ten metres long, attached to the collar via a good quality swivel. To help you gauge how far the liner has travelled underground, mark the line with a waterproof ink mark every metre. Never mark the line by tying knots in it, as these cause problems by constantly snagging on every underground obstacle.

Clothing

Clothing used for ferreting needs to be robust, strong and able to protect the

Carrying rabbits – using a piece of baling twine to hock the rabbits and hang over one shoulder is an easy way to carry home small numbers of conies.

wearer from the rigours of the weather and natural obstacles such as thorns. All clothing must be roomy enough to allow free movement, without causing discomfort, and outer garments are best if windproof and preferably waterproof.

Clothes worn during a ferreting expedition have a nasty habit of becoming very wet, dirty and smelly, and so it is advisable to have a complete change of clothing and footwear in the car, along with a couple of towels for drying yourself. A large plastic bag will be useful to place dirty clothing in.

*   *   *

Do not make the mistake of believing that an accumulation of equipment will help you be better at ferreting; many of the best ferreters use only a minimal amount of gear. Knowledge and experience, neither of which can be bought, are the two essentials for good sport and effective rabbit management.

Chapter 8

Showing, Racing and Pets

Although ferrets have been used for thousands of years for hunting purposes, many of today's owners prefer to have them as 'fancy' animals, i.e. for exhibition purposes, for shows or simply as pets. Decried by some, welcomed with tremendous enthusiasm by others, these alternative uses for ferrets have made the animal even more popular and, in some countries, are the only reasons for possessing ferrets, as hunting with them is either banned or incredibly difficult to obtain permission for.

THE FANCY

Ferret shows are today popular around the world, and it has proved impossible for me to discover when the first 'official' ferret show took place. However, I know that back in the 1970s small shows were regularly being staged in the UK, as I was often asked to judge at them. They were small, low-key affairs, and those taking part looked on them as harmless fun. Today, with so many people having showing as their *raison d'être,* some take the subject very seriously indeed. There is nothing wrong with this, as long as the well-being of the ferrets is always put first, and any hobby needs to be fun or it soon becomes work.

Throughout the world, people have sat down and described their ideas of the perfect ferret; this is usually very subjective, rather than objective, and varies enormously between countries, regions, clubs and even individuals. If you are intent on being successful at showing, you should carefully study the standards for your area, and your animals must comply with them if you are to stand a chance of winning.

My worry in all of this is that the ferret fancy may go the way of the dog fancy; the standards for many dogs are exaggerated, and the animals themselves are often put at physical risk because of this. Some breeds of dog are bred with heads so large that they cannot give birth normally, and have to endure an elective Caesarean section; others are bred with such short noses that they have difficulty breathing. Some standards call for animals with such fine bones that, when the animal behaves at all naturally, it is in danger of breaking those bones, while others have so much wrinkled flesh on their faces that their eyesight suffers.

To my mind, we should set standards for the ferret which encourage breeders to produce animals which are entirely normal – capable of full working lives, and able to exist without any interference from man, if need be. This would seem to be common sense, but as we have seen, this seems to be sadly lacking in many fancies. Even today, many in the ferret fancy are breeding animals with exaggerated physical characteristics. I have witnessed many ferrets – both hobs and jills – which are twice the size of the normal animal; their heads are so big and broad that I am sure some complications must occur at birth. Please let common sense prevail, for the ferret's sake.

I have here set out a few principles that I hope clubs will base their standards on; I have not set hard and fast rules, simply common-sense considerations, which are in the best interests of the ferret. The terminology used in fancy animals also needs a little explanation, and again I have given a few guidelines.

The categories under which points are allotted are:

Colour and markings
Type
Fur
Size
Condition
Eyes and ears

The first category, colour and markings, is the one which has most variance, and so I will not say anything on this except to suggest that, in patterned varieties fifty per cent of the points for this category should be allocated for colour and markings and fifty per cent for pattern. I also feel that for shows to be successful, different colours should be judged in separate classes, with the overall results coming from the result of the comparison of each animal against its standard, and *not* each other.

'Type', or configuration as it is sometimes called, refers to the overall shape of the ferret. The ferret should have a long, lean body, which arches when the animal is at rest, with short legs; its gait should be easy and effortless. The tail should be about half as long as the body. The head should be proportionate to the body, with the hob having a much broader skull; the face of the animal may be pointed or blunt, as neither is detrimental to the animal. The head should be well set into the body, the profile showing a smooth curve from nose over head to nape of neck. The neck should be cylindrical, and noticeably long, blending gradually with the slender and elongated thorax and abdomen.

The fur should be soft and very dense, with due allowances made for animals which have been neutered (this tends to give them a better coat). The coat should consist of fine underfur and longer, more coarse guard hairs, and the tail should be well furred.

The ferret should be lean, not fat, and of a size which would allow the animal to lead a normal predatory life. Allowance should be made for sex, i.e. hobs are larger than jills.

The ferret should be fit, curious when well-awake, and tame to handle. The flesh should be firm with no surplus fat, and the coat have a healthy sheen, with the ferret being quite clean.

The eyes should be small and widely set with the ears being small and discreet, set well apart on the head.

The teeth should be white and the ferret should have a complete set. The jaws should meet correctly.

Penalties should be incurred for the following reasons:

Intractability
Disease or injury
Excess fat

Sores, wounds or scars
Dirty fur
Missing limb, eye, ear, foot, toe or tail
Any deformity believed by the judge to be hereditary
Under/over-shot jaws
Damaged or missing teeth

No penalties should be incurred by ferrets with a slight amount of wax in their ears. This is perfectly normal and is actually an aid to the ferret's well-being. Where the ears have been scrubbed out, and may even show signs of irritation from the process, penalties should be incurred.

FERRET RACING

This is a sport that started in the USA and has proved popular in many countries, particularly the UK, where such events feature strongly at most country fairs and similar events. When properly organised, the crowds find the spectacle fascinating, and the ferrets seem to enjoy themselves.

The legend surrounding the invention of this sport is that oil workers used ferrets to pass lines through pipelines; once through, the lines could be attached to cameras, welding machinery, etc., which could then be hauled through the pipe. At night, with very little to do, the workers would take bets on the fastest ferret to go through a length of pipe.

When the idea arrived in the UK, it was suitably tweaked to allow for British conditions. Today, the average ferret race takes place between several ferrets, which have to run down identical lengths of piping (all of the same bore). At intervals along the pipes, gaps are left (to help the public see where the ferrets are), and the trick is to encourage the ferret to ignore them and continue to the end of the piping.

Again, the potential problems are that some breeders will develop lines of ferrets which are very small and fast, to enable the animals to complete the course at high speed; the danger is that these animals will be *too* small and fast to allow them to lead a predatory life, i.e. they will become unnatural.

* * *

Whatever the sport, while the ferrets are waiting to be judged or raced, the organisers should arrange suitable holding facilities, where the ferrets can be rested, fed, watered and exercised, and that area should have shelter from prevailing temperatures and weather conditions. (See Appendix 6).

PETS

As a pet, the ferret leaves most of its competitors in the shade. Despite the unwarranted bad reputation that the ferret has for being aggressive and biting needlessly and indiscriminately, ferrets are easily tameable, and with regular, careful handling can be handled with total safety by all but the smallest child.

Their size means that they can take a reasonable amount of fairly rough handling, and can also be taken for walks on a lead, as a small dog or cat would be. Guaranteed to turn heads as they walk down the street on a harness and leash, the ferrets really seem to enjoy their airing and the chance to see so many happenings that they would not normally witness.

Many people keep ferrets in the house, giving them complete freedom to wander around; when questioned on this, the owners simply state that it is no different from a cat or dog having free rein in the human home. Ferrets do, however, have the knack of getting into all the wrong places. I have heard of ferrets which nest in the bottoms of cookers, in drawers of underwear, and even in human beds.

If your ferret is allowed free access around the home, ensure that all visitors are made aware of this, and watch where everyone puts their feet, since a booted foot can cause serious injury to a ferret. Every room should be equipped with a litter tray, which will need to be emptied and cleaned regularly, and food and water dishes should also be routinely serviced. Don't forget that ferrets like to hide their food around their home, and so you shouldn't be surprised if you find a half-eaten chunk of meat tucked away inside your favourite shoe.

TRANSPORT

Whatever your reason for having ferrets, you will need some sort of carrier to transport the animals, whether it be to a show, a race, the vet, or simply on your holiday. The choice of a suitable carrier should be given some thought and different types will appeal to different people.

Perhaps the most popular type of carrier is that sold in pet shops for cats and other small animals. Made from high-impact plastic or polycarbonate, they have a door which swings open at the front, and a carrying handle on top. They are also made in such a way as to allow the top to be completely separated from the bottom, and this makes cleaning easy.

Many people use wooden boxes, and these are adequate if one remembers that wood is porous, and so is difficult to keep scrupulously clean, and needs to have regular attention to prevent the ferret's urine soaking into the wood with the resultant smell and rotting. Regular treatment of the box with a safe wood preserver will help give the box a long working life.

Some ferret owners like baskets; I do not. They are insanitary, being impossible to keep clean, and draughty for the animals.

Whatever type of container you use for transporting your ferrets, ensure that they have adequate ventilation, and never leave them in the sun, or in a closed car, as the heat can rise to a lethal temperature very quickly.

BATHING

Although not generally in favour of bathing ferrets on a regular basis, I recognise that there are times when a ferret requires a bath. It may have become soiled while being transported, it may be ill or have an infestation of ectoparasites, or you may be ensuring that it looks its best before appearing on the show bench.

Shampoos designed for use on cats are ideal for ferrets; use of an insecticidal shampoo, even if your ferret does not have fleas, will act as a preventative measure and is to be recommended.

Use a bowl specifically for the ferret; using the same bowl that you wash your clothes or even dinner pots in is extremely insanitary and a crass habit. Run a small amount (about 4–5cm) of tepid water into the bottom of the bowl, and place a mat or piece of cloth in it. This will help the ferret to obtain a purchase, preventing it slipping around, and therefore panicking. Next, carefully place the ferret in the bowl, talking softly to it the whole time.

When the ferret is reasonably happy, remembering that repetition will aid acceptance of this, gently pour some of the tepid water over it, working it into the fur all the way down to the base; avoid the eyes and ears. Next, pour a little shampoo on the fur and gently massage it well into the coat, again avoiding the eyes and ears. When the ferret is covered with a good lather, and the shampoo has been worked well down into the coat, rinse with fresh tepid water. Repeat this procedure, ensuring that the ferret's coat is well rinsed of all remains of the shampoo.

Lift the ferret gently from the bowl and wrap in a towel; again this should be one used solely for the ferret. Rub vigorously, removing most of the water, and then place it in an open-fronted cage out of draughts. Position a hair-dryer, or electric fan heater, about twenty centimetres from the front of the cage, and switch it on to low, directing the stream of warm air into the cage. Keep a careful check on the ferret and, within about five to ten minutes, it should be dry and can be removed from the cage. Never return a damp ferret to an unheated cage area, or one that is out-of-doors, regardless of the ambient temperature or prevailing weather conditions.

If the ears are very dirty, dip a cotton bud into warmed olive oil, and gently wipe the pinnae (ear flaps) one at a time, using a separate cotton bud for each ear. Once completed, use a fresh, dry bud to remove the debris. *Never* push the bud into the ear, nor should you use undue force or a scrubbing motion.

If you fear that your ferret has ear mites, consult your veterinary surgeon, who will prescribe specific ear-drops.

Chapter 9

Ailments and First Aid

Properly cared for, the ferret is an extremely healthy and hardy animal, and many owners have never experienced serious medical problems with any of their ferrets. However, accidents can and do happen, and illness is certainly not unknown among ferrets. Unfortunately, ferrets are still very much 'wild animals' and as such, tend to mask symptoms of illness – often until it is too late to help them. Any unusual behaviour among your ferrets should have you checking more vigilantly, and no veterinary surgeon will mind being consulted for an animal which is not at death's door.

In many countries, ferrets are so cheap that vets rarely see any in their surgery; it is cheaper and easier for uncaring owners to kill the luckless ferret than to have it treated. As more people are attracted to ferrets, and more knowledge is gained about all aspects of the animal, this sad state of affairs will hopefully come to an end.

THE BASICS OF FIRST AID

The old saying that things are as easy as A B C is very true when it comes to first aid, the ABC being – AIRWAYS, BREATHING, CIRCULATION. In other words, if the patient is not breathing, get the airways clear before you try to get the animal breathing, and only once the animal is breathing should you concern yourself with the heart and/or bleeding.

All first aid principles are the same, regardless of species involved, and I cannot recommend too highly that everyone should have basic training in the subject. Such organisations as the Red Cross and the St John Ambulance Service run courses at very reasonable rates, and going on one may help you save a life.

I have found two widely used definitions which I feel help to explain the subject:

'First aid is the skilled application of accepted principles of treatment, on the occurrence of an accident, or in the case of a sudden illness, using facilities and materials available at the time. '

'First aid is the approved method of treating a casualty until they are placed, if necessary, in the care of a qualified medical practitioner, or removed to a hospital.'

The objects of first aid are three-fold – to sustain life, to prevent the patient's condition worsening and to promote the patient's recovery. The methods used are:

a)   assess the situation;
b)   diagnose the condition;
c)   treat immediately and adequately.

Remember to ensure your own safety at all times.

Prevention is, of course, better than cure. Ensure that your ferrets are given a good, balanced diet; their cages cleaned regularly; they are not subjected to extremes of temperatures; are kept away from draughts, and are not stressed unnecessarily. Any cuts, abrasions or bites must be cleaned and treated immediately.

First Aid Kit

Some of the most common ferret injuries – minor cuts and abrasions – occur when the animal is going about its daily routine, or is being worked, and it is important that they are treated as soon as possible in order to minimise adverse effects on the animal. In order to do this, a small first aid kit should be taken along on every hunting foray and, of course, you should have the necessary skill and experience to treat these minor injuries; if there is any doubt as to the seriousness of the injuries, or the ferret's general condition, veterinary treatment should be sought as soon as practical.

A suitable first aid kit for ferrets should contain the following items as a minimum, and should always accompany you on any field trip and be near at hand at all times.

1. *Nail clippers*

These should be top-quality and can be used for trimming the ferret's nails. Use the type which work on the guillotine principle, where one blade hits the other, rather than on the scissor principle, which can result in nails being pulled out. Ferrets' nails should not be cut too short; leave about 5mm beyond the quick.

2. *Tweezers*

For the removal of foreign bodies.

3. *Scissors*

These should be curved and round-ended and are to be used to cut off the fur around any wound. They must *not* be used for trimming nails.

4. *Antiseptic lotion*

For cleansing cuts, wounds and abrasions.

5. *Antihistamine spray*

Occasionally, a ferret will be stung by a bee, wasp or other such insect which has managed to find its way into the ferret's cage. The sting should be removed and antihistamine applied.

6. *Cotton wool*

Used for cleaning wounds, cuts and abrasions and for stemming the flow of blood.

7. *Surgical gauze*

Used for padding wounds and stemming the flow of blood.

8. *Adhesive plasters*

Although these will soon be chewed off by the ferret, they are useful for applying directly to small wounds and for keeping dressings in place. They can also be used for minor splinting.

9. *Bandages*

A selection of small bandages should be kept for binding broken limbs and wounds. They will, of course, be temporary as the ferret will chew them off.

10. *Cotton buds*

Ideal for cleaning wounds and the application of ointments, etc. With care, they can also be used to clean the pinnae (ear flaps) but *under no circumstances* should you poke these (or any other object) down the ear canal.

11. *Table salt*

A solution of table salt (two teaspoons of salt to 0.5 litres of water) is a good solution to wash debris from wounds and to counter infection.

12. *Sodium bicarbonate*

On a wet compress, this will help reduce swelling.

13. *Alcohol (surgical spirit)*

Useful for the removal of ticks, etc.

14. *Styptic pencil*

To help stem the flow of blood.

Many of the items listed can be used on either ferret *or* human injuries.

COMMON DISEASES AND AILMENTS

The following is a list, in alphabetical order, of some of the most common diseases and ailments which may afflict ferrets. It is not intended to be a 'do-it-yourself guide to medicine', merely to help owners recognise any potentially dangerous symptoms in their ferrets. I have also included information on preventative measures which should be taken to avoid medical problems. In all cases of ill or injured ferrets, I strongly recommend that you seek the advice of a veterinary surgeon; do not delay this, as such time lapses can often lead to a sick ferret going beyond the point where a vet can save the animal. Time waited is time wasted.

Aleutian Disease (AD)

This disease is caused by a parvovirus and is an immune deficiency disease. It takes its name from the Aleutian strain of mink in which it was first discovered in the USA in 1956, although it is believed that there is now a specific ferret strain of this disease. Since 1956, cases have been recorded in Canada and New Zealand. The first cases reported in the UK were in early 1990, probably via mink imported from the USA. No true connection has ever been proved between these imported mink and AD in ferrets, but vets and Defra officials still retain their suspicions that this is the case.

Aleutian disease is described as being transmitted both vertically and horizontally, i.e. it can pass from parents to offspring and also from unrelated ferrets to other ferrets; this latter type of transmission occurs through body fluids. These can be saliva, blood, urine, faeces and aerosol droplets from the lungs, via sneezing, cough-ing and even simply breathing. This aerosol transmission can occur up to a metre away from the infected animal. It is obvious that hobs (males) should not be shared in breeding programmes, as this is an almost guaranteed method of transmitting the disease; neither should hoblets (vasectomised males) be shared between kennels of ferrets. (Hoblets are used to bring the jill out of oestrus, without getting her pregnant – a vital operation if the jill is not to succumb to many of the ailments that prolonged oestrus can cause in ferrets; see Oestrogen-induced Anaemia). The numbers of animals reported as infected with AD are tiny.

Symptoms vary tremendously and may include black tarry faeces, weight loss, aggressiveness, recurrent fevers, thyroiditis, posterior paralysis, and eventual death; in stressed animals, death may occur suddenly. There is no specific treatment, but antibiotics and steroids may give temporary relief. Diagnosis of this disease in a living ferret can be confirmed by serology, the necessary blood being obtained by clipping the toenails in such a manner that, using a capillary tube, a vet can obtain blood from the animal. With dead animals, a post-mortem examination, which should always be conducted in the case of a sudden or unexplained death, or where a disease is suspected of causing the death, will be necessary to confirm the diagnosis.

Alopecia (Hair Loss)

This is often caused by the feeding of too many raw eggs, which contain a compound which inhibits biotin, resulting in this condition. There are also many other causes of hair loss in ferrets, including seasonal environmental changes, and it is recommended that any ferrets manifesting such symptoms be given a thorough examination by a vet. (See also Mites.)

Hair loss on the tail can occur in either sex and at any age, ranging from a very small area of the tail to the compete tail, and in most cases, the naked skin is seen to have black spots on it; these are blocked pores – 'blackheads'. Usually, though not always, such hair loss occurs when the animal is in its moult of late summer or early autumn and may last for up to three to four months; the condition is also often to be seen when the spring moult occurs. Not all ferrets in the same cage will be affected, and even affected ferrets are not necessarily affected every year.

Veterinary tests in many affected animals have not shown any reason or causal agents, and it is believed that the condition is simply a case of excessive moult. If in doubt, consult your vet.

Abscesses

Abscesses are simply wounds which have filled with pus, the bacteria *Staphylococcus* and *Streptococcus* being the usual cause. They can be caused by a variety of occur-rences, such as bites, cuts or damage to the inside of the mouth caused by bones in the ferret's diet. In order to prevent wounds developing into abscesses, ensure that they are thoroughly cleaned and disinfected. Once abscesses have developed, they will require lancing and draining, often several times, and it may be necessary for the animal to be given a course of broad-spectrum antibiotics, such as ampicillin. Obviously, action of this type must only be taken by a veterinary surgeon.

Actinomycosis

An acute hard swelling of the ferret's neck, actinomycosis is probably caused by abrasions to the animal's oesophagus, usually by feeding too many day-old chicks in the diet. Affected ferrets will be listless, anorexic and have a fever (a temperature of up to 40.8 °C is not uncommon). Refer the animal to a vet who will probably administer injections of cephalosporin; you will then have to keep the ferret nourished on a liquid diet until full recovery is achieved.

Bites and Stings

These need to be split into four categories – insect, snake, rat or ferret.

Insect Bites (Including Stings)

Clip a little fur away from the area so that you can actually see the problem, then wash with saline solution. Bees leave their sting in the victim; wasps do not. If there is a sting present, it should be carefully removed with tweezers and then the area wiped with cotton wool (or a cotton bud) soaked in alcohol, such as surgical spirit. For wasp stings, a little vinegar will prove beneficial, while for bee stings use a little bicarbonate of soda. Dry the area thoroughly, and use antihistamine or apply a wet compress to help reduce the irritation and swelling.

If the ferret has been bitten or stung in the throat, veterinary attention must be sought as soon as possible; such stings can cause swelling that may block the airways and thus kill the ferret.

Rat Bites

Rat bites are one of the most dangerous of all bites that a working ferret may suffer. Rats can carry many harmful diseases and so it is essential that no risks are taken.

Clip away the fur from around the wound, ensuring that the clippings do not become entangled in the wound itself. Wetting the scissors is recommended, as the hairs stick to the blades rather than falling onto the wound; dipping the scissors into a jug of water after each snip will remove all hairs from the metal. Thoroughly clean the area of the wound with a saline solution and then an antiseptic liquid; dry and apply liberal amounts of antiseptic dusting powder. If the wound is large, or you have good reason to believe that the rat was infected, take the injured ferret to the vet as soon as possible after the injury, where the vet may well administer an injection of antibiotics. I would always recommend this action after any rat bite – better safe than sorry.

Snake Bites

There are many venomous snakes and, although it is unusual for ferrets or dogs to be bitten by these reptiles, it does sometimes happen. During the spring or early summer the snakes are rather lethargic, especially the gravid (pregnant) females. At such time, they will keep still as long as possible, even when approached. If your ferret or dog does not see the snake and stands on it, the snake will bite. It is most important that you keep the injured animal as calm as possible (and you must also remain calm as your actions will influence the animal), and seek immediate medical attention.

Probably the most common injuries to ferrets are caused by other ferrets. This jill has had her neck bitten by the hob during mating.

Ferret Bites

Although it may at first seem strange, ferrets are more likely to suffer from bites from other ferrets than from any other animal, especially during the breeding season, when the hob takes hold of the jill by the scruff, hanging on tightly and, very often, breaking the jill's skin with his teeth. These types of injuries are not usually serious, provided that they are given first aid treatment as soon as possible.

The area of the bite must be clipped of fur, and the wound thoroughly washed with a saline solution followed by an antiseptic liquid. A good dusting with an antiseptic wound powder will finish the job. If action is not taken, the wound may fester and result in abscesses.

Bleeding

NEVER apply tourniquets (see Cuts).

Botulism

Botulism is a killer. The disease is caused by one of the most common bacteria known to science, *Clostridium botulinum,* usually 'Type C', a natural contaminant of most wild bird cadavers. When this bacterium comes into contact with any decaying flesh (i.e. meat), it causes a deadly toxin to be formed. If this flesh is then eaten by an animal, the toxin affects its victim by attacking the animal's nervous system causing paralysis, at first usually in the hind legs. Eventually, this paralysis will affect the body's vital organs, and leads inevitably to the death of the affected animal. There is no cure or treatment for this, and ferrets are among the most susceptible animals to botulism.

In order to try to prevent this deadly disease, pay particular attention to the meat that you feed; defrost frozen meat and feed immediately. If there is any doubt whatsoever about the meat, boil it for at least fifteen minutes before feeding. Botulism is not contagious and sometimes only one animal of a group may succumb to the illness. If, for whatever reason, you believe that your ferrets have a high risk of this disease, it is possible to have an annual toxoid injection to provide some protection for your ferrets.

Breathing Problems

Ferrets gasping for breath are obviously showing symptoms of some form of breathing difficulty; this may be heatstroke, fluid on the lungs or an obstruction of some kind. Many obstructions can be removed from a ferret's mouth with a cotton bud or even a finger.

Artificial respiration, though difficult, is possible with ferrets. If a ferret has stopped breathing, rather than give mouth-to-mouth respiration, hold the ferret by its hind legs and, keeping your arms straight, swing the animal to left and then to right. This transfers the weight of the ferret's internal organs on and off the diaphragm, causing the lungs to fill with and empty of air. Keep this up until the ferret begins breathing on its own, help arrives, or you believe the ferret to be beyond help.

Convulsions

Convulsions are a symptom, an indication that the ferret has an infection of some kind or has been poisoned, and *not* a disease. There are obviously many possible causes for convulsions, and one of the most common in captive ferrets is heatstroke, or the 'sweats'. However, if your ferret is suffering from convulsions, you should seek urgent medical attention for the ailing animal.

Cuts, Abrasions and Other Wounds

There are several different types of wound which a ferret may suffer, and each requires a slightly different technique.

Incised (Clean) Cut

These are straight cuts, as one would get from a sharp knife blade; as such, they bleed profusely. This bleeding, which can be very frightening to people not used to such things – a little blood appears as 'gallons' to most lay people, even though there may be only a thimble full – helps clean the wound of debris, and this lessens the possibility of infection.

Bleeding should be stemmed by direct pressure, if at all possible; where it is not, apply indirect pressure on an artery at the heart side of the wound. Elevating the injury will enable gravity to help reduce the blood flow. Apply a suitable dressing; large and/or deep cuts will almost certainly require sutures (stitches) from a vet.

Lacerated Cuts

These are tears in the skin, as caused by barbed wire, for instance, and will bleed less profusely than incised cuts. The big danger is that the injury will have pushed dirt and debris into the wound, and the lack of bleeding will mean that the dirt is not washed out. You must clean the wound with a saline solution (salt water – two teaspoonfuls to 0.5 litres of warm water); dry it and apply dressing if necessary.

Contusion (Bruise)

This is a sign of internal bleeding, and a careful watch must be kept on the injured animal. If shock sets in, seek veterinary advice immediately.

Puncture (Stab)

Puncture wounds, which can be caused by nails, slivers of wood and other such objects, usually appear very small at the surface but, of course, could be very deep. *Never* remove any object from a wound, as this may aggravate the injury and/or allow large amounts of bleeding. Apply pressure around the wound site, using a dressing to maintain the pressure, and seek medical attention immediately.

Gunshot Wounds

If you work your ferrets, then it is quite possible that they may be accidentally shot. The most common type of wound to a working ferret is from a shotgun, in which case the animal will be peppered with balls of shot, which will all require removal; bleeding should not be profuse. The other type of wound is from a solid projectile, i.e. a bullet. Bullets make two wounds – one in and one out. Where 'sporting' (hollow point) ammunition is used, the exit wound will be several times larger than the entry wound.

In all cases of gunshot wound, stem the flow of blood, keep the animal calm, checking for signs of trauma, and seek veterinary treatment immediately.

Dental Problems

Ferrets, like humans, have teeth, and also, like humans, occasionally have dental problems. They occasionally damage their teeth, either while working or even in their cage when chewing the wire mesh; gingivitis, a gum disorder, is also quite common in ferrets. The build-up of food debris often leads to dental problems, and the diet of the ferret is, therefore, an important factor in the condition of the teeth. Any problems with a ferret's teeth must be treated by a vet.

Diarrhoea

Diarrhoea is a symptom and *not* a disease; it is indicative of a problem, which may be serious or minor, but will still require investigation. In ferrets this condition is often referred to as the 'scours', and is usually, but of course not always, a sign that the animal has been fed on a poor diet, or its food is contaminated. Feeding milk sops, too many raw eggs, too much fat, a sudden and abrupt change of diet, and food which has gone off, all have the effect of scouring a ferret. Diarrhoea can also be indicative of some other, more serious affliction, such as poisoning, internal parasites or even stress.

If you are feeding your ferrets a proper balanced diet of fresh, or freshly defrosted, cadavers, they should not suffer from loose motions as a matter of course. If you have been feeding them on a diet of bread and milk, their motions will almost always be soft (often liquid), light brown in colour, and extremely smelly. If fed on a complete dry diet their motions will be solid, almost black, and have little smell. Keep a record of everything you feed them, particularly anything out of the ordinary.

Diarrhoea causes the animal to dehydrate, and this can lead to irreparable body damage (particularly of the kidneys), and even death. You should isolate the affected ferret(s) and keep it/them on a water and electrolyte (a solution of essential minerals) regime for twenty-four hours, dosing with kaolin solution about every two hours. After the fast, food intake should be gradually built up again; do *not* put the animals straight back on their original diet, otherwise the whole problem may recur. Chicken, rabbit and fish are excellent 'invalid' foods, and ideal for this task.

If the diarrhoea persists, or if there is blood in the motions, a vet must be consulted immediately.

Distemper, Canine (CD) (Hard Pad)

Ferrets are highly susceptible to canine distemper, a virus which is one of the most common fatal diseases in ferrets; dogs are the most common source of such an infection, with the incubation period being between seven and nine days. In some countries, owners have their animals vaccinated against this disease as a matter of course, while in others this does not happen. Indeed, in some countries there is no distemper vaccine licensed for use in ferrets; the use of an unsuitable vaccine may have fatal results. If you are considering this preventative course of action, consult your vet.

Symptoms of the disease are swollen feet, leading to hard pad, the actual thickening of the soles of the feet and a classical sign of distemper infection; runny eyes and nose, diarrhoea, lack of appetite, a larger than average thirst and a rash, usually under the chin. In its latter stages, the infected animal will vomit, have convulsions and, shortly before dying, will pass into a coma.

This disease is highly contagious and, at the first signs of distemper, all infected ferrets must be isolated. Ensure that you thoroughly disinfect your hands after handling a sick animal; as the name suggests, dogs can also contract this disease, and it is all too easy to spread it. Immediate veterinary advice must be sought, although only very mild cases can be treated; it is often kinder, both to the infected ferrets and the others in your care, to have the infected ferret put down once the diagnosis is confirmed by a vet.

Enteritis

Enteritis is an inflammation of the intestines, causing diarrhoea, and is very common among kits and young stock. If your ferrets are experiencing diarrhoea, and show signs of blood in their faeces, this may indicate this condition. It can be caused by different things but, usually, it is the bacterium *Escherichia coli,* often referred to by its abbreviated name of *E.coli,* and formerly known to science as *Bacillus coli.*

Immediate treatment with a broad-spectrum antibiotic, supplemented with regular doses of kaolin, may cure this condition. If left untreated, the affected animal will most definitely die.

Another major cause of enteritis is *Campylobacter,* in humans, this type of 'food poisoning' is known as dysentery. The most effective antibiotics for use against this are chloramphenicol and gentamicin.

As mentioned under the heading of 'Diarrhoea', affected animals must be given large amounts of water and electrolyte to avoid the dangers of dehydration.

Foot-rot

This condition was extremely common until a few years ago, and in some areas is still a major problem. It is usually caused solely by keeping ferrets in dirty, wet conditions, allowing a mite, *Sarcoptes scabiei,* to infect the animals. The symptoms of foot-rot are swollen, scabby feet and, if left untreated, the claws will eventually drop off. Affected animals must be isolated immediately, and all other ferrets – especially those kept in the same cage – examined. All bedding and wood shavings must be removed from the cage and burned, with the cage itself being thoroughly disinfected with a suitable disinfectant.

Veterinary treatment must be sought for all infected ferrets, as home treatment is only very rarely effective; time waited is time wasted.

Fractures

Fractures are caused by either direct or indirect pressure on the bones, which may crack or actually break. Where the bone is broken and pierces the skin, this is known as an open or compound fracture, and all others as closed fractures. Signs of such injury are obvious – painful movement of the limb, tenderness, swelling, loss of control of the limb, deformity of the limb, unnatural movement of the limb, and crepitus (the sensation or, in very bad cases, the sound of the two ends of the bones grinding on each other).

Keep the patient quiet and steady and support the injured limb, immobilising it with bandages and splints if necessary, to prevent it moving and causing greater damage. Raising the limb will help reduce discomfort and swelling (by reducing the blood flow).

Hard Pad see Distemper

Heatstroke (the 'sweats')

Ferrets cannot tolerate high temperatures, reacting adversely to too much heat, and may well die from heatstroke, often referred to as 'the sweats', a rather misleading term since ferrets cannot sweat. As in all things, prevention is better than cure, and the siting of the cage (as discussed in Chapter 3) is very important, as is the position that they are left in while out working. In the confines of a carrying box or even a motor car, the temperature can quickly rise to a dangerous level, even in the cooler sunshine of autumn and spring. No animals should be left unattended in a vehicle, or transported in such a manner that they or their carrying box are in full sunlight. Remember, the sun does not stay in the same position throughout the day, and that, even if the box or car is in the shadows when you leave it, it may not stay that way for long. When you return, your ferrets may well be dead.

In extreme summer temperatures, or in areas where it is known that temperatures will be high, every effort must be made to insulate the cage, and place it in an area where it and its inmates are protected from the full effect of varying temperatures. Where it is not possible to keep the ferrets' cage as cool as one would like, wet cloths may be hung over the cage to keep the temperature down, although they will soon dry out, and so require constant attention throughout the day. Placing bricks on each corner of the cage roof, and then positioning a piece of timber over them, will act as 'double glazing', and will be extremely effective in reducing heat build-up in the cage.

The first sign of heatstroke or heat exhaustion is an agitated ferret in obvious distress. If in their cage, affected ferrets will stretch out and pant heavily; if left untreated, they will eventually collapse, pass into a coma and die.

Immediately a ferret shows symptoms of heatstroke, you must act – and fast; delay can be fatal. The ferret's body is overheating, and so your first task must be to lower its body temperature. With mildly affected ferrets, simply moving them to a cool area and ensuring a steady passage of cool air over them is usually effective; a light spraying with cold water from a plant mister is beneficial. In bad cases, where it is literally make or break, the best method is to immerse the animal to the neck in a bucket of cold water, repeating this procedure regularly for the next few minutes, by which time the ferret should be showing signs of recovery. Ensure that the ferret is thoroughly dried, and placed in a cage in a cool area, with a small amount of hay for bedding; this hay will also help to dry the ferret's coat. This is a very drastic 'treatment', and should only be attempted where the animal is very badly affected.

In all cases, it is vital to keep the head cool, as brain death can occur – the brain is quite literally 'cooked'. Veterinary advice should be sought at the earliest opportunity.

Hypocalcaemia

This is caused by a lack of calcium in the blood. It can occur three to four weeks after the jill has given birth. Posterior paralysis and convulsions are common symptoms, while the cause is usually poor diet; feeding a diet heavily dependent on day-old chicks will almost certainly lead to this complaint. Consult your veterinary surgeon immediately, who will probably administer an intraperitoneal injection of calcium borogluconate, which gives a speedy response in affected animals. After this injection, a calcium-rich diet is essential for total recovery.

Influenza

Ferrets are susceptible to human influenza, and can catch influenza ('the 'flu') from their owners, or vice versa.

Symptoms in ferrets are the same as in humans – fever, sneezing, lack of appetite, listlessness, runny eyes and a nasal discharge. In adult ferrets the condition is not usually serious, with most ferrets making a spontaneous recovery, although antibiotics may be necessary to control some secondary infections. Influenza is, however, almost always fatal in young ferrets.

Infected ferrets must be isolated to prevent the spread of this condition. The wearing of a mask by humans will help prevent the crossing of the infection from ferret to human, and vice versa.

Leptospirosis
The ferret has a natural resistance to this disease, often called 'rat catcher's yellows' or 'Weil's disease', and is highly unlikely to contract it. Consequently, vaccinations for leptospirosis are unnecessary and should not be used.

Leukaemia
This disease is often found in unmated jills, but is almost completely undetectable until the ferret is almost at death's door. The presence of any 'lumps' in the animal's groin or abdomen, or under the jaw, may indicate this disease. Medical advice should be sought immediately. See also 'Oestrogen-induced Anaemia'.

Mange see 'Mites'

Mastitis
This is the inflammation of the jill's mammary glands, and often occurs when the jill is in the early stages of feeding young. It is a very painful condition that requires immediate medical attention. The glands become very swollen and hard and the kits can obtain very little, if any, milk. Unless it is cleared up quickly, the kits will probably die and the jill will, at best, be very ill. The disease is caused by infection by *E. coli* and treatment usually consists of antibiotics, preferably ampicillin.

Mites
Mites have unpleasant effects on the ferret and there are three types commonly found on the animal.

Sarcoptes scabiei causes two types of mange in ferrets, which can become infected through coming into direct contact with other infected animals, such as rodents, or simply by being on infected ground. One type of mange (sarcoptic) causes alopecia and pruritus (an intense itching), while the other causes only foot or toe problems ('foot-rot'). The first sign of mange is persistent scratching, even though there is no obvious cause such as fleas. Eventually, the skin will become very red and sore, a symptom that is easier to notice in albino ferrets than in ones with polecat coloration. As the disease progresses, these sores cause baldness and the sores become even worse.

A parasiticidal wash (e.g. bromocyclen) must be applied to the affected areas, or injections of ivermectin administered; this drug cannot be used in the first month of pregnancy, or it will cause congenital defects. The cage must be thoroughly treated, soaking it in a strong solution of disinfectant or bleach, which must be washed off before any ferret is returned to the cage.

Be warned – mange can be contracted by humans, when it is known as scabies.

Ear mites, *Otodectes cynotis,* are common in ferrets and can easily be treated with ear drops which contain gamma BHC or with ivermectin injections. If your ferret seems to spend a lot of time scratching its ears, an investigation is called for. A build-up of wax in the ears, dotted with black specks, is a sure indication that ferrets have ear mites; the black specks are probably spots of dried blood, and the ear mites are usually white or colourless and are not visible to the naked eye – a magnifying lens or otoscope being required. If left untreated, the irritation caused by these mites will cause the ferret to scratch, sometimes until its ears actually bleed. The mites can move down the aural canal and infect the middle ear; such an infection will cause the affected animal to lose its sense of balance. This may be indicated either by the ferret simply being unable to hold its head straight or, in more serious cases, by constantly falling over.

It is important that all ferrets that have been in contact with the infected ferret are also treated, as ear mites can infect other animals who may not show any symptoms for some time.

Seek veterinary advice in all cases of ear mite, loss of balance, etc.

The harvest mite, *Trombicula autumnalis,* can cause sores on the underside of the neck and trunk of ferrets, particularly during the autumn period; a wash in a bromo-cyclen-based product will be effective.

Oestrogen-induced Anaemia

When a jill is not mated the levels of oestrogen, the female sex hormone, in her body will rise and can have serious effects on the health of the ferret, often causing progressive depression of the bone marrow and sometimes resulting in a condition known as pancytopoenia – the abnormal depression of all three elements of blood – which is debilitating and potentially fatal. Signs include weight loss, alopecia, anorexia, pale lips and gums, difficulty in breathing and, in later stages, darkening of faeces (caused by blood), and bleeding sores on the animal's flanks and abdomen. There may also be secondary infections.

In its advanced stages, treatment is highly unlikely to be effective, but in early stages, spaying or hormonal treatment to stop oestrus may be used, and repeated transfusions of fresh whole blood containing 1ml of sodium citrate can be used effectively.

Again, prevention is better than cure, and jills which it is not wished to breed from should be neutered (spayed), or given drugs (e.g. proligestone), or served with a hoblet (a vasectomised male ferret). The latter will almost certainly result in a 'phantom pregnancy' (pseudo-pregnancy) which will last for the full forty-two day term, after which the jill will return to oestrus, and need to be served again.

Osteodystrophy

Defective bone formations, often a result of hyperphosphorosis (too much phosphorous in the diet) are caused by feeding a diet consisting entirely of, or rich in, muscle meat; this lacks calcium, and leads to a deficiency of this vital mineral. The problem usually manifests itself in young ferrets (six to twelve weeks) which have difficulty walking, moving instead with the gait of a seal, with the legs (particularly the front legs) sticking out to the side of the body rather than pointing to the floor; death is common. As the bones of such animals are soft and deformed, the ferrets never completely recover, and will always have deformed legs and spine, making movement awkward, difficult and painful. A top-quality vitamin and mineral supplement can help, but it is better to prevent the problem by feeding a diet of complete cadavers rather than 'meat'. Veterinary advice must, of course, be sought whenever ferrets have difficulty walking in a normal manner.

Parasites

Even the most pampered ferrets can suffer from the unwanted attentions of parasites, either internal (endoparasites) or external (ectoparasites). Ferrets can suffer from worms, *Toxocara* or *Toxascaris*, but these are not usually problematical. A vet will prescribe an anthelmintic (an agent which is destructive to worms) such as mebendazole or fenbenzadole.

Heart worms, coccidia, *Toxiplasma* and *Pneumocystis* are sometimes found. Veterinary examination, and clinical investigation of the faeces, will be necessary to positively identify the actual problem.

The first signs of a worm infestation are an insatiable appetite coupled with a steady loss of weight. Sometimes, segments of the worms may be found in the ferret's faeces before other symptoms indicate a problem. Do *not* attempt to treat ferrets with powders and doses that your local pet shop may recommend for a dog or cat. Seek veterinary advice.

Fleas – *Ctenocephalides*, and ticks – *Ixodes ricinus*, are external parasites that most working ferrets will catch at some stage, especially if used for hunting. These ectoparasites are contracted from other animals such as dogs, cats, rabbits and other such species. Insecticidal preparations intended for dogs and cats are usually safe to use on ferrets. With new preparations coming on to the market all the time, care must always be taken to ensure that any product is safe *before* you use it on your ferrets. Do not forget that you will have to adjust the dosage to suit the size of the ferret, as per the manufacturer's instructions.

All of the bedding and wood shavings from the ferret's cage must be removed and burned and the cage thoroughly disinfected. Use the powder as a prophylactic (preventative measure) to treat all shavings, bedding and even the cage itself. Do *not* use such powders and sprays in any cage where a jill is still feeding her kits, as there is a danger of poisoning the litter.

Ticks are rather more difficult to deal with than fleas, but they do respond to some sprays and powders, although in some countries these may only be available from veterinary surgeons. Care must be taken to ensure that the mouth parts of ticks are completely removed from the ferret's skin, otherwise infection and abscesses can occur; never simply pull ticks out. Paint alcohol on the tick using a fine paintbrush, and the tick should have died and dropped off within twenty-four hours; if not, simply repeat the process.

Although some authorities suggest that ticks be burned off with a lighted cigarette, this should never be attempted. It is all too easy to burn the ferret with the cigarette and the alcohol method is much more effective, with none of the dangers.

There are now a number of devices on the market which have been specifically designed to remove ticks, owing mainly to concern regarding Lyme's disease in humans. These are a type of sprung forceps which grip the tick around the head; the device is then gently twisted backwards and forwards, until the tick comes out. I have tested several such implements and have had complete success with each; I now always keep one in my first aid kit.

Posterior Paralysis ('The Staggers')

The causes of paralysis in ferrets are many and varied, including disease of the spinal disks, hypocalcaemia (lack of calcium in the blood), Aleutian disease, viral myelitis (inflammation of the spinal cord), cancer of the spine, vertebral trauma, or even a dietary deficiency. It is often caused by injury and paralysis can also be inherited. If the latter is the case, the disorder will show itself in young litters. There is usually no cure for this and it is, therefore, best to have the animals destroyed. The parents of such litters should not be used for breeding again. In all cases of paralysis, veterinary advice must be sought immediately.

Pregnancy Toxaemia

Leading to the sudden death of the jill just before she is due to give birth, the cause of this problem is currently unknown but is probably linked to a poor diet.

Pyometra

This is the accumulation of pus within the uterus and is only occasionally seen in ferrets; when it does occur, it is immediately after the start of a pseudo-pregnancy. Organisms responsible for this condition in ferrets include *Streptococcus, Staphylococcus, E. coli* and *Corynebacterium.*

Affected ferrets will be anorexic, lethargic, and will often have a fever. Medical attention must be sought immediately, as the uterus may rupture, causing peritonitis, and an ovariohysterectomy will be required urgently.

Rabies

In countries where this disease is prevalent, ferrets can become infected. Clinical signs of the disease include lethargy, posterior paralysis and anxiety; recovery of infected ferrets is not unusual.

Ringworm (Dermatophytosis)

Although rare, this condition is occasionally seen in ferrets. It is nearly always contracted by contact with an infected cat or, extremely rarely, dog. Caused by the fungus *Microsporum canis,* this disease is not caused by a worm, as many are led to believe by its name. The condition manifests itself with hair loss and bald, scaly patches of skin. This disease is a fungal infection. Home treatment is sometimes successful, but it is strongly recommended that veterinary advice is sought. As the condition is transmissible to man, it must be treated immediately (see also Mites).

Sarcoptic Mange see Mites

Scours see Diarrhoea

Shock

Shock, which is an acute fall in blood pressure, is often evident after a ferret has been involved in an accident or has been injured; certain diseases can also cause this condition. It manifests itself with cool skin, pale lips and gums (owing to the lack of circulation); faint, rapid pulse; staring but unseeing eyes.

The victim must be kept warm and the blood circulation returned to normal as soon as possible. Massaging the ferret will help circulation, and wrapping it in a towel or blanket will help keep it warm. The affected animal should be kept quiet and warm, and veterinary treatment sought as soon as possible.

Skin Tumours (Neoplasia)

Common in ferrets, in all cases surgery is necessary and should be carried out as soon as possible. These tumours are, however, very difficult to eliminate and frequently recur. Warts also occur on ferrets and so it is vital that histopathology is used to confirm all diagnoses. Consult your vet if any animal develops lumps anywhere on its body.

'The Staggers' see 'Posterior Paralysis'

Thiamine Deficiency

Where ferrets are fed a diet consisting of a large proportion of day-old chicks, raw fish or eggs (or a combination of any or all of these items), a thiamine deficiency is almost guaranteed. Symptoms include anorexia, lethargy, weakness of the hindquarters, and convulsions. Injections of the vitamin B complex usually have a rapid effect, and recovery is usually total. Avoid this problem by supplying your ferrets with a proper diet. All cases of convulsions, paralysis or unusual behaviour should be referred to your vet.

Tuberculosis

Ferrets are susceptible to avian, bovine and human tuberculosis, with symptoms which include paralysis of the limbs, diarrhoea and wasting of the body. Almost always fatal, the disease is highly contagious; contact your vet at once if you suspect this condition in your ferrets.

Urolithiasis (Gall Stones)

Varying in size from a particle of sand to quite a large stone, gall stones sometimes occur in ferrets. Treatment consists of antibiotics, surgery and the use of special diets to dissolve the stone.

Zinc Toxicity

Ferrets cannot tolerate high levels of zinc and may become ill through the use of galvanised feeding dishes or even by licking cage bars. Symptoms include anaemia, lethargy and weakness of the hind legs; liver and kidney failure can quickly follow and so all suspected cases should be referred to a vet as soon as possible, although there is no actual treatment and affected animals are unlikely to recover. It is common for vets to advise euthanasia for affected ferrets to avoid undue suffering.

*   *   *

In many cases health problems are avoidable, and the best way to ensure healthy ferrets is to indulge in good husbandry, including the provision of an adequate balanced diet. Ferrets have the disturbing habit of often not showing any symptoms until they are almost dead. If you suspect any of your stock of 'sickening', contact your veterinary surgeon without delay; time waited is time wasted.

Chapter 10

Genetics

Many people fight shy of genetics, thinking that the subject must be incredibly difficult and that to understand it, one will require a degree in zoology; this is simply not true. Once certain basic rules and principles have been learned, a few technical terms committed to memory, and a little common sense applied, the subject will be seen to be extremely logical and quite easy. Too many people, both in the ferret fancy and other fancies, will not make this effort, preferring to fumble along, perhaps asking 'experts' for advice from time to time. However, if you do not have a good (i.e. working) knowledge of genetics, your breeding programme will never be more than moderately successful in the long term. At best, you will produce ferrets of average quality, at worst you will produce ferrets that cannot be passed on, and yet are useless to yourself.

BASICS

Within the scope of this book, I will only attempt to guide the reader through the basics of the subject. However, once you have grasped those basics, life will be easier and, if you wish, you will be in a position to take your studies further. For more detailed information, refer to my book *Ferret Breeding*.

It is necessary first to understand that *every* animal receives a half of its inherited traits from each parent. Sometimes, for reasons that we shall discuss later, it may seem that some offspring inherit more qualities and characteristics from one parent than from the other, but this is not so.

The inherited components are known as genes, which are carried on the chromosomes, rather like beads on a string. It is the job of the genes to tell the cells of the body how to behave – to change shape, to grow or to change colour, for instance. In the ferret, there are forty (i.e. twenty pairs) of chromosomes. Along this string of beads are points known as 'loci' and, normally, each locus is occupied by the same gene.

SEX CHROMOSOMES

One pair of chromosomes determine the sex of the individual, and are logically known as the sex chromosomes (all the others are known as autosomes). The male chromosome is referred to as Y, while the female chromosome is known as X. A female ferret will have two X chromosomes (XX), and the male will have one X and one Y chromosome (XY). Obviously then, it is the male's chromosomes which influence the sex of any progeny, and the resultant ferret has a fifty-fifty chance of being male, and exactly the same chance of being female.

At fertilisation, the germ cells (the female egg and the male sperm) contain twenty single chromosomes each (owing to a process called meiosis) so that, on fertilisation, the resultant zygote contains 20 pairs again, as will every cell (except the germ cells) of every individual ferret.

Once fertilised, the egg (now known as a zygote) grows and develops into a young ferret. This growth is accomplished by progressive division of the cell, the number of cells doubling at every stage i.e. 2, 4, 8, 16, 32, etc., until the resultant tissue takes on the appearance of an identifiable foetus. When the cells divide, the chromosomes are 'shuffled', so that the genes which accompany them can form new, as well as established, combinations. It is not known exactly how many genes are carried on a chromosome, but some authorities believe that it could be tens of thousands. Such large numbers are necessary to control all of the physiological functions of the body, as each can only act in a very simple manner, and so vast quantities of them are required to control body functions.

MUTATIONS

Obviously, the constant creation of new cells means that the chromosomes (and hence the genes) are being constantly duplicated. These duplications usually go according to plan, and the copy made is an exact copy. However, very rarely (about once in a million copies), a slight copying error may occur. This means that a duplicated gene is slightly different from its predecessor. Remember that genes control the functions and appearance of an individual, and so a changed gene could result in a noticeable difference in the animal concerned, e.g. coat colour, type of fur, etc.

Such an abrupt change, where it results in a change in appearance of the individual, is referred to as a 'sport', 'mutation', or 'mutant'. As far as we, as ferret breeders are concerned, the mutations that would most interest us are those which result in either a change in appearance, or have a detrimental effect on the ferrets. When a mutation occurs, thus creating a 'mutant allele', this mutant allele will occupy the same locus as its normal cousin did. During cell division (i.e. growth), some genes may 'crossover' from one chromosome to the other chromosome. Obviously, genes which are close together on the chromosome will be less likely to crossover than those that are far apart. Two genes on the same chromosome are said to show 'linkage', and the degree of linkage is an indication of the distance between the genes, i.e. the closer together, the stronger the linkage.

MENDELIAN GENETICS

Gregor Mendel was a member of the Augustinian monastery in Brünn, in Austria (now known as Brno, in the Czech Republic). In 1856, he carried out a vast number of experiments on hereditary factors, choosing the garden pea *(Pisum sativum)* as his subject. This plant has a number of easily identifiable characteristics – wrinkled and smooth seeds; long and short stems; red and white flowers, etc. – and he carefully cross-pollinated these and meticulously recorded the results over many years. From these results Mendel drew several conclusions, and these form the foundations for today's understanding of heredity. He described his work in an obscure natural history journal in 1866. It wasn't until 1900, however, that the scientific world became aware of Mendel's findings, but today 'Mendelian Genetics' form the foundation of all of the work that has been, and is still being, carried out on heredity.

All chromosomes, and therefore all genes, are present twice in the individual, but only once in the germ cells. At its simplest, genetics concerns the passing on from one generation to the next of several characteristics. In order to grasp the principles involved, we must take the subject one step at a time. We will, therefore, look at the way in which a single characteristic is passed on. This is known as 'monohybrid inheritance'.

MONOHYBRID INHERITANCE

Let us take an example of a pair of *pure* (i.e. true-bred) polecat-coloured ferrets mating together. As all of the inherited material passed on to the next generation must be the same (as regards colour and markings, i.e. polecat), then all of the young must also inherit these characteristics. When these ferrets breed with each other, again they can only inherit the same colour and markings – polecat.

Suppose, however, that we mated a pure polecat-coloured ferret with a pure albino.

It is obvious that, given that the progeny from any mating will always inherit fifty per cent of its traits from each parent, the litter will all be carrying genes for both polecat *and* albino (the first litter in a series of litters bred for genetic purposes is always referred to as Fl; the second as F2; the third as F3, etc, the 'F' standing for 'filial'). However, some genes dominate others; these are known as dominant genes, while those which are dominated, are known as recessive. Recessive genes will only show themselves when the animal has two helpings of that recessive gene, i.e. is pure-bred. This can only happen when *both* parents possess this gene. The polecat gene, as it is the original (natural or normal) colour is always dominant to all other colours. Therefore, when the two come together, although the genetic make-up (the genotype) of the ferrets is both polecat and albino, the physical appearance (the phenotype) is that of a polecat, as this gene effectively masks the action of the albino gene.

As in most jobs, practitioners use their own form of shorthand, and the geneticist is certainly no exception. All *mutant* genes are given a letter; dominant genes are signified by using an upper case (capital) letter, while reces-

A true polecat hob belonging to the author.

sive genes are given a lower case letter. If an animal carrying a mutant gene is mated to an animal *not* carrying that gene, then the shorthand used to designate the ferret not carrying the mutant gene would be the reverse of that for the ferret that *is* carrying that gene, i.e. the shorthand for the gene that produces the albino is 'cc'; if such an animal were to be mated with a pure-bred polecat-coloured ferret, that is obviously *not* carrying that recessive gene, the genotype of the polecat-coloured ferret would be given as 'CC'.

In our example, the pure polecat-coloured ferrets would be designated as 'CC' (showing that both sets of genes are the same and that albino is recessive to polecat), while that of the albino is assigned the code 'cc'. Geneticists also use their own method to simplify the way that calculations are carried out. The principle is exactly the same as that used by mathematicians when working out problems involving numbers. The method is known as the 'punnett square', and involves the letters representing each set of genes being written along one side and the top of the square. These are then added together to give the answers in each square.

Male Genes (Pure Polecat)

		P	P
Female Genes	P	PP	PP
(Pure Polecat)	P	PP	PP

Male Genes (Pure Polecat)

		C	C
Female Genes	c	Cc	Cc
(Pure Albino)	c	Cc	Cc

Where a dominant gene (i.e. one denoted by upper case letters) is present, it will be this that dictates the phenotype of the ferret. Therefore, in both of the foregoing examples, all progeny would be phenotypically polecat. The resultant young of the mating of two pure polecat-coloured ferrets have identical genes (i.e. P and P) and are said to be homozygous for polecat coloration; but the progeny of the mating between the polecat-coloured ferret and the albino have a gene from the polecat (C) and one from the albino (c), and are said to be heterozygous (or 'split') for albino.

If these splits are now self-mated (i.e. mated brother to sister), we will see that the recessive genes will now have a chance to show themselves.

Male Genes (splits)

		C	c
Female Genes	C	CC	Cc
(Splits)	c	Cc	cc

Obviously, those ferrets carrying the 'cc' genotype will have the phenotype of the albino while those carrying 'CC' and 'Cc' will have a polecat-type phenotype. In all such matings (i.e. between two splits both with the same genotype) *approximately* twenty-five per cent of the litter will show the recessive gene in their appearance. This is often expressed as '25:50:25', or '1:2:1' to show the proportions of different geno-types that one would *theoretically* expect to get in the litter. However, it should be noted that any percentages given in genetics are based on the same principles as statistics, and therefore only approximate; to achieve these types of figures, huge numbers must be bred. This phenomenon will never be more apparent than in a litter from a mating that only produces a single kit; obviously, whichever phenotype that kit displayed, the actual percentages would be one hundred per cent of one and zero per cent of the other(s).

If one mated a split with its recessive parent, the litter would contain fifty per cent splits (i.e. phenotypically polecat) and fifty per cent pure albinos.

Another way of illustrating inheritance involving two ferrets of different geno-types is as follows:

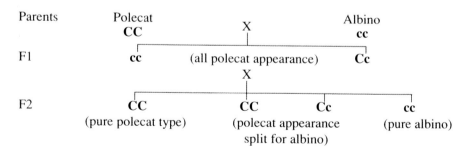

The principles outlined above are very important for breeding good stock and 'crossing' to improve a specific line. For example, if you have a line of polecat-coloured ferrets where all of the animals are of good type and size, but have a line of albinos that do not possess these qualities, you can mate one variety to the other, knowing that, within two generations, you would again be producing albino animals.

DI-HYBRID INHERITANCE

The story is complicated when *both* parents are splits – but not for the same colours. The calculations to solve this problem are carried out in exactly the same way as previous calculations, i.e. with the use of a punnett square. No matter how experienced you may get, you will never be wasting your time by drawing a punnett square and working out all of the possibilities on paper, rather than relying totally on mental arithmetic.

Although very little work has been carried out on ferret genetics, or at least on the coat colour inheritance, we know from the results of the breeding of other species that some colours and varieties cannot be produced by 'mixing' others; it is necessary to have one or more animals which are actually carrying the specific mutant gene. Other varieties, however, are obtained from a mix and consequent combination of genes.

Many people speak glibly of 'polecat' and 'albino' colorations in ferrets; however, if these were the only two genotypes, we would not be able to produce all of the other varieties that are so common today. The exception to this statement is the dark-eyed white. Every dark-eyed white that I have seen has been the result of careful breeding (using inbreeding) to eliminate all dark hairs in the 'blue' or 'silver' ferret, leaving an almost white animal with dark eyes. I am told that some breeders have found the mutant gene which creates a true black-eyed white, but I have never seen such an animal. I have, however, seen white ferrets with very dark (ruby) eyes.

Geneticists have discovered many genes which produce a 'white' animal, and only one gene that produces the true albino. Remember that the term albino means an animal completely lacking in pigmentation, i.e. it has no colour whatsoever; its fur is pure white, and its eyes are opaque, although the blood seen through the eyes gives them a pink appearance. What then of the 'yellow' ferrets? While some will tell you that these are merely the result of albinos being kept in dirty conditions (and with some animals this may be true), careful examination of these yellow ferrets will reveal that the fur has its own colour, and the eyes are not opaque, but pink. When crossed with a polecat-coloured ferret, many of the offspring appear to be normal polecat-coloured animals.

However, if one looks at these very closely, certain differences appear. Some of the guard hairs are now 'white'; certain points of the body (e.g. feet, chest and top of head) are also 'white'. The most important difference, however, is slightly more difficult to see; the eyes have a dark red tinge to them, and this is known to geneticists as a ruby eye. The coloration will only show under certain lighting, and this is why many owners have never noticed the colour difference. Such an animal is known by many different terms – silver sable, white-mitt, sandy, etc.

It is the presence of these 'other' white genes which appears to be giving the ferret world such a profusion of varieties at present. I am, however, certain that, as more ferrets are bred, more mutant alleles will be found, and the number of varieties will increase dramatically.

SEX-LINKED INHERITANCE

It will be remembered that it is the male ferret that has both an X and a Y chromosome (XY), while the female has two X chromosomes (XX). If a mutation occurs on

NEW COMPLETE GUIDE TO FERRETS

either of these two sex chromosomes, it is referred to as 'sex-linked'.

In most species, the most obvious example of such a gene is the one which is responsible for both the tortoiseshell and the yellow varieties – To – and it extends the yellow pigment in the fur. In a homozygous animal (ToTo), this will result in a yellow coat, while in a heterozygous (Toto), it will give rise to a tortoiseshell variety. This occurs because of the presence of one normal X chromosome which will limit the effect of that chromosome carrying the To gene, and the yellow colouring will therefore be restricted. For this reason, normally only female tortoiseshells will ever appear, but both male and female yellows will appear. *Very* occasionally, a male tortoiseshell will appear, but these are invariably infertile.

LETHAL AND SEMI-LETHAL GENES

In some mutations, if the foetus has a double helping of that particular gene, i.e. they are homozygous, the animal dies. These are known as lethal genes. It is not known if any such genes occur in the ferret, although the likelihood is that they do. When these are seen, it can safely be assumed that the animals are heterozygous and, when either of these two varieties are bred together, there will be a one-in-four chance of any of the resultant foetuses having a homozygous combination of these genes; they will, therefore, die while in the womb, being reabsorbed. The siblings that survive will not, please note, suffer any adverse effects from the lethal gene. They will also all be splits, as it is obviously never possible to pure-breed these varieties.

RESEARCH

We all need to discover more about the ferret's genetic make-up; one of my biggest disappointments while researching this book was my failure to find many scientific and valid results on the genetics of the ferret. This was not because of my shortcomings – many though they are – but owing to the lack of such material.

I am sure that you, like me, will find it absolutely amazing that, after having kept ferrets in captivity for so many years, we have obtained such a small amount of information on such a basic subject. Even those individuals and companies who have (and still are) breeding ferrets for their pelts – and would obviously benefit from scientific facts regarding coat colour inheritance – appear to have failed to keep proper records, or at least to communicate the results of their breeding programmes. It is a sad state of affairs, and one that should be addressed as soon as possible by all with a real interest in the ferret – for whatever reason.

You, the reader, can help, simply by keeping accurate records of all breeding carried out. A central body could easily collate this data, and reach conclusions which will help us all. The UK's National Ferret School has agreed to act in this manner, and all interested parties should contact them for further details. (See Appendix 4).

142

Appendix 1

Ferret Facts and Figures

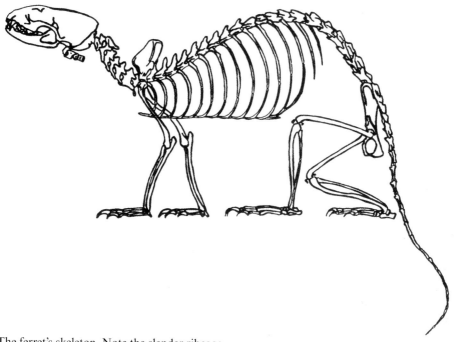

The ferret's skeleton. Note the slender ribcage.

Average weight (adult)	Between 400g (♀) and 1.5kg (♂)
Average size (adult)	Between 35 and 60cm (hob up to twice size of jill)
Average lifespan (in captivity)	Between eight and twelve years
Age at puberty	250 days (approx.)
Age at sexual maturity	Eight to twelve months (spring after birth)
Normal breeding season (Britain)	Early March to late September
Number of chromosomes	Forty (twenty pairs)
Breeding season	Triggered by photoperiod – requires longer days than nights
First possible mating	Six months
Duration of oestrus	Ceases when mated; if unmated, will continue to end of 'season', causing potential ill-health and medical problems

Signs of oestrus	Vulva swells, becomes vivid pink colour, with secretion
Signs of breeding condition in male	Testes swell and descend into scrotum
Duration of oestrus	Until mated or end of spring/summer (can last for up to six months if not mated)
Duration of mating	Several hours (dependent on photoperiod)
Ovulation	About thirty hours after coitus
Reduction of vulva	Begins seven to ten days after mating, complete within two to three weeks
Palpation	Ten days after vulva totally reduced
Gestation (pregnancy)	Forty to forty-four days
Number of young (in one litter)	One to fifteen (average six to eight; fifteen largest recorded)
Litters per year	Under normal conditions, one or two; with extra lighting (i.e. adjustment of photoperiod) up to four
Post-natal oestrus	One to two weeks after weaning
Weight at birth	Five to fifteen grams
State at birth	Altricial – i.e. blind, deaf, naked and entirely dependent upon the mother
Ears open	Thirty-two days
Eyes open	Thirty-two to thirty-five days
Fur	In dark-coloured kits, starts to appear within five to seven days; good covering by four weeks
Deciduous teeth	Begin to erupt at about fourteen days; all showing by eighteen days
Permanent canine teeth	Forty-seven to fifty-two days
Shedding of deciduous canine teeth	Fifty-six to seventy days
Movement	Kits as young as two to three weeks will manage to crawl out of nest (to be dragged back by the mother)
Weaning	Six to seven weeks
Weight at weaning	300–500g
Age attain adult weight/size	Four to five months
Rectal temperature	38.6 °C (Range 37.8–40 °C)
Heart rate	220–250 bpm
Respiratory rate	30–40 per minute
Number of toes	Five per foot
Teeth	Incisors 3/3, canines 1/1, premolars 3/3, molars 1/2
Nipples	Two rows of four (eight in total) on both sexes
Vertebrae	C7, T14, L6, S3, COL14–18, 14 pairs of ribs
Other anatomy	No appendix, caecum, or male prostate gland; under-developed sweat glands

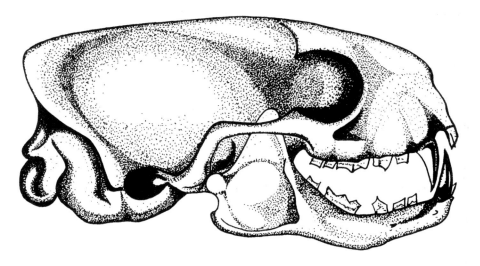

A ferret's skull, clearly showing the large canine teeth for which the animal is renowned.

Sexing

The ano-genital distance of the female is half that of the male

Appendix 2

Glossary of Ferret and Ferreting Terms

Albino A white ferret, with no pigmentation in its flesh or eyes. The eyes are opaque but, owing to the fact that blood can be seen behind the eyes, they appear pink. There are several other types of white ferret which are *not* albinos.

Austringer The traditional term for one who keeps and hunts hawks.

Autosomes Chromosomes other than those which determine the sex of the progeny of a mating.

Bedding The material supplied to ferrets for their nest. Top-quality meadow hay is best; do not use cotton wool, newspaper or any long-fibred synthetic material.

Benching The act of placing ferrets on the show-bench, for the attention of the judge officiating at that show.

Business The collective term for a group of ferrets.

Cage The container in which ferrets are kept. These can be commercially manufactured or home-made by breeders using wood, plastic or metal. (See also Court and Cub)

Canines The front teeth of a ferret which it uses to kill its prey.

Carbohydrates Food constituents which provide the body with nutrition and material for growth.

Cannibalism The killing and eating of one ferret by another ferret.

Carrying box A container designed to hold one or more ferrets, ideal for travelling or transporting the ferrets to and from a show, race or working area. Sometimes called a travelling box.

Chromosomes Thread-like bodies, carried inside the nucleus of every germ-cell, which carry the genes of the individual animal.

Classes	At a ferret show, entries are divided into classes, to facilitate judging. These classes are usually dictated by colour, variety or sex of the ferret concerned.
Coitus	The physical act of mating.
Collar	A collar is placed around the neck of the liner ferret and a line attached to it. Pet ferrets are often put on to a collar and lead and taken for walks.
Cope	A metal muzzle. (See also Muzzle)
Court	A traditional type of cage, rather like an aviary, in which ferrets are kept.
Colony	Term used to describe the keeping together of a group of ferrets together, usually for commercial breeding purposes, where one male is kept with several females.
Crepuscular	The term used to describe an animal that is active around the hours of dusk and dawn.
Cub	The traditional name for the hutch-like cage in which some ferrets are kept.
Culling	The removal and killing of some kits in a litter which is thought to be too large for the mother to rear. This is totally unnecessary, as the female ferret will know her own limits, culling the litter as she sees fit, starting with the weakest members.
Dam	The mother of a litter.
Dominant	A gene which will always show itself in the phenotype of a ferret, even if there is only one of these genes present. Such genes are denoted by the use of an upper case (capital) letter.
Ectoparasite	A parasite which lives on the outside of the host animal e.g. fleas, lice and ticks. (See also Parasite and Endoparasite)
Electronic detector	A device for tracking ferrets (and terriers), used while ferrets are working underground. The device consists of a collar-mounted transmitter (worn by the ferret or terrier) and a hand-held receiver. Both are pre-tuned. A modern ferret detector can be used to a depth of almost 5 metres.
Endoparasite	A parasite that lives inside the host animal (e.g. in the

intestines, etc.) such as tapeworms. (See also Parasite and Ectoparasite)

Fats — Constituents of food which provide the ferret's body with energy. A surplus of fats in a ferret's diet will cause it to become overweight, with danger to its health. This is even more so if the ferret is also deprived of sufficient exercise.

Ferret — The domesticated European polecat (*Mustella putorius*). The term is often (erroneously) used to describe only the albino version of this species.

Ferret detector — See Electronic detector

Fibre — The indigestible material present in some foods which helps stimulate the action of the intestines. Used to be known as 'roughage'.

Fitch — The ferret's fur.

Fitchet — A term used originally to describe the young of a *wild* polecat and a domesticated ferret, but today it is used for any polecat-coloured ferret.

Foot-rot — A disease caused by mites and resulting in the feet of the afflicted ferret quite literally rotting. Once thought to be caused solely by the ferrets being kept in dirty and wet conditions.

Gene — A hereditary factor of inherited material. Genes are carried on the chromosome.

Genetics — The study of the ways in which certain characteristics are passed on from one generation to the next.

Genitals (Genitalia) — The external sex organs of a ferret.

Genotype — The genetic make-up of a ferret.

Genus — A group of animals containing species which closely resemble each other, e.g. the weasel tribe (*Mustelidae*).

Germ-cell — The egg of a female and the sperm of the male ferret.

Gestation — Pregnancy. In ferrets, this is between forty and forty-four days, usually forty-two.

Greyhound ferret	A tiny, agile ferret. Sometimes referred to as a 'whippet ferret'.
Heat	A term often used to describe oestrus.
Heatstroke	The effect of too much warmth on a ferret. If this condition is not treated, it can be fatal. Sometimes referred to as 'the sweats'.
Hob	A male ferret or polecat.
Hobble	A castrated hob.
Hoblet	A vasectomised hob.
Hospital cage	A special cage for sick or injured ferrets. It is fitted with a thermostatically controlled heater.
Hybrid	A cross-bred ferret, i.e. heterozygous. The term usually refers to the offspring of a wild polecat and a domesticated ferret.
Hybrid vigour	The increased vigour and resistance to disease often found in the offspring resulting from the mating of completely unrelated ferrets.
Inbreeding	The practice of breeding together very closely related ferrets. (See also Line-breeding)
Incinerator	A device for burning rubbish (especially soiled shavings and bedding). Essential for the safe disposal of bedding and shavings, etc. from the cages of infected and ill ferrets.
Incisors	The front teeth of a ferret.
Inheritance	The manner in which certain characteristics are passed from one generation to the next.
Intractability	Lack of tameness; impossible to handle without the risk of being bitten.
Jill	A female ferret or polecat.
Kit (or Kitten)	A young ferret; usually used to describe a ferret of sixteen weeks of age or less (the same terms are also used to describe young rabbits). Comes from the name of the young of the polecat, i.e. pole-kitten.
Life expectancy	see Longevity.

Line	A 'family' of ferrets, bred for several generations.
Line-breeding	A moderate form of inbreeding.
Litter	The young ferrets produced at one birth.
Locus	Position on a chromosome occupied by a specific gene (plural loci).
Longevity	Length of life. In the ferret, this is about eight to twelve years in captivity. The oldest ferret that I have ever known was 'Izzy', a poley jill, who was almost fourteen years of age at the time of her death.
Mask	The markings on the face of a polecat (or polecat-coloured ferret). This mask is usually darker in the summer than in the winter.
Mating	See Coitus.
Minerals	Minute constituents of a ferret's diet, without which it will not have a balanced diet, with a consequent adverse effect on health.
Monohybrid inheritance	The inheritance of a single characteristic.
Musk	The foul-smelling scent produced by the anal glands of ferrets and polecats.
Mutant or Mutant allele	The changed gene which results in a change in an animal – usually, its appearance.
Muzzle	A device for preventing the ferret from biting (either humans or its natural quarry). Made from leather, string or metal; they must *never* be used on ferrets while they are working. Also the name for the ferret's nose.
Myxomatosis	A deadly viral disease amongst rabbits. Thought to have been deliberately introduced to the UK from France, to 'control' the numbers of wild rabbits.
Nocturnal	The term used to describe an animal that is active by night, sleeping by day.
Oestrus	The state in which a jill will accept a mating. (See also Heat)

Oestrous cycle	The sexual cycle of a female ferret.
Ovulation	The release of eggs into the womb to be fertilised by the hob's sperm.
Parasites	Animals which live on or in other animals (hosts) in a manner which is detrimental to the host. Includes worms, fleas, mites, lice and ticks. (See also Endoparasites and Ectoparasites)
Paunch	(Verb) To remove the 'guts' of a rabbit. (Noun) An animal's stomach.
Pen	An old-fashioned term for a cage.
Phenotype	The physical appearance of a ferret.
Photoperiodism	The dependence on the daytime/night-time (or simply light and dark) ratio of various biological functions, particularly the commencement of oestrus.
Polecat	The common name of the animal *Mustela putorius*. Strictly, this name should only be used to describe the wild polecat, but today it is also commonly used to describe *any* domesticated ferret with polecat type markings. The name originates from the French *poulet chat* – chicken cat.
Poley	A domesticated ferret with the typical wild polecat markings.
Pregnancy	See Gestation.
Proteins	The basic constituents of all living things and an essential constituent of a balanced diet; essential for growth and tissue maintenance.
Quiet	The term used to describe an easy-to-handle ferret.
Recessive	A gene which is masked unless another identical gene is present in a ferret. Such genes are denoted by writing their code in lower case letters.
Records	All of the information concerning your ferrets. Such records must always be kept and, of course, kept accurately.
Roughage	An old-fashioned term for fibre.

Sandy	A coloured ferret with colouring between an albino and a polecat.
Scours	Diarrhoea.
Season	See Oestrus.
Sex chromosomes	The chromosomes responsible for determining the sex of a ferret. Males have one X chromosome and one Y. Females have a pair of X chromosomes. All other chromosomes are known as autosomes.
Sex-linked	The term used to describe a characteristic where the mutant allele is carried on the sex chromosome and is, therefore, governed by the sex of the individual.
Sexual dimorphism	The differences exhibited between the sexes, e.g. the male ferret always has the capacity to grow larger than the female (up to twice the size).
Sibling	Brother or sister; litter mate.
Sire	The father of a litter.
Skulk	When a ferret refuses to leave the mouth of a tunnel or its nest-box/cage, it is said to be 'skulking'.
Sops	Bread and milk, sometimes (though wrongly) fed to ferrets.
Split	The term used to describe a ferret whose parents were of different colours.
Sport	The term used to indicate a ferret that is genetically different from the norm.
Stud	An individual ferret 'kennel', where ferrets are bred. Some clubs allow one to register a stud prefix (name) which is exclusive to the registrant. Often used to describe the male ferret used for breeding purposes.
Stud-book	The record (not necessarily a book) of all ferrets in a particular stud.
Sweats	See Heatstroke.
Terrarium	Cage.

Travelling box	See Carrying box.
Variety	A specific colour and coat type of the ferret, e.g. silver sable.
Vaccination	An injection of a mild form of a specific pathogenic micro-organism, which causes the body to form antibodies, thus helping to prevent the treated animal from acquiring a full dose of the specific disease, e.g. distemper.
Virus	Organism able to cause disease.
Vitamin deficiency	The lack of certain important vitamins; this term is usually used to indicate the result of such a deficiency, rather than the actual lack of the vitamin(s) in question.
Vitamins	Organic compounds, essential to the health of a ferret. Usually referred to by letters of the alphabet, e.g. A, B, C, etc.
Water bottle	The bottle, equipped with a stainless steel spout, in which a ferret is supplied with water. Every ferret must have access to clear, clean, fresh water at all times.
Weaning	The development of the eating habits of ferret kits when they progress from being dependent upon their mothers for food, to being capable of feeding themselves, i.e. eating solid food.
Zoonose or Zoonotic disease	Disease capable of being transmitted from animals to humans, e.g. salmonellosis.
Zygote	A fertilised egg.

Appendix 3

Bibliography

The inclusion of a title does not imply any endorsement by the author; neither does the omission of a title imply lack of approval.

ABWAK, 'Management of Canids and Mustelids'. Proceedings of the 5th symposium of the Association of British Wild Animal Keepers (ABWAK) (1980)

Birks, J.D.S., and Kitchener, A.C. (eds), *The Distribution and Status of the Polecat Mustela putorius in Britain in the 1990s*, The Vincent Wildlife Trust (1999)

Birks, Johnny, *Mink*, The Mammal Society (1986)

Blood, D.C., and Studder, V.P. (eds), *Baillières's Comprehensive Veterinary Dictionary*, Baillière Tindall (1988)

Bourdon, Richard M., *Understanding Animal Breeding*, Prentice Hall (1997)

Clark, Tim W., *Conservation Biology of the Black-Footed Ferret*, Wildlife Preservation Trust International (1989)

Clutton-Brock, Juliet, *A Natural History of Domesticated Animals*, Cambridge University Press and the British Museum (Natural History) (1987)

Cooper, B. and Lane, D.R. (eds), *Veterinary Nursing*, Pergamon (1994)

Cooper, J.E., Hutchinson, M.F., Jackson, O.F. and Maurice, R.J. (eds), *Manual of Exotic Pets*, The British Small Animal Veterinary Association (1992)

Corbet, G.B. and Hill, J.E., *World List of Mammalian Species*, The British Museum (Natural History) (1980)

Corbet, G.B., and Harris S. (eds), *The Handbook of British Mammals* (3rd edn), Blackwell Scientific Publications (1991)

Dallas, Sue, *Animal Biology and Care*, Blackwell Science (2000)

Dalton, Clive, *An Introduction to Practical Animal Breeding* (2nd edn), Collins (1985)

Day, Christopher, *The Homoeopathic Treatment of Small Animals*, The CW Daniel Company Limited, 1990

Drakeford, Jackie, *Rabbit Control*, Swan Hill Press (2002)

Everitt, N., *Ferrets – their Management in Health and Disease*, N. Everitt (1897)

Fortunati, Piero, *First Aid for Animals*, Sidgwick & Jackson (1989)

Fox, James G., *Biology and Diseases of the Ferret* (2nd edn), Lea & Febiger (1998)

Grzimek, B. (ed), *Grzimek's Animal Life Encyclopaedia* (vols 10, 11 and 12), Van Nostrand Reinhold (1972)

Hammond, K., Graser, H-U., and McDonald, C.A., *Animal Breeding – The Modern Approach, Post Graduate Foundation in Veterinary Science*, University of Sydney (1992)

Hodson, Anna, *Genetics*, Bloomsbury (1992)

Honacki, J.H. and Kinman, K.E., *Mammal Species of the World*, Allen Press (1982)

Jurd, Richard D., *Instant Notes in Animal Biology*, BIOS (1997)

King, Carolyn, *Weasels and Stoats*, Christopher Helm (1989)

Lloyd, Maggie, *Ferrets – Health, Husbandry and Diseases*, Blackwell Science (1999)

MacDonald, David, *The Velvet Claw*, BBC Books (1992)

MacDonald, David (ed), *The Encyclopaedia of Mammals Volume 1*, Guild (1985)

Marchington, John, *Pugs and Drummers*, Faber & Faber (1978)

Mason, I.L. (ed), *The Evolution of Domesticated Animals*, Longman (1984)

Matthews, L. H., *The Life of Mammals* (vols 1 and 2), Weidenfield and Nicolson (1971)

McKay, James, *Ferret Breeding*, Swan Hill Press (2006)

– *Practical Falconry*, Swan Hill Press (2002)

– *The Ferret & Ferreting Handbook*, The Crowood Press (1989)

Morris, Desmond, *The Mammals*, Hodder and Stoughton (1965)

Pardiso, Nowak, *Walker's Mammals of the World*, Johns Hopkins University Press (1991)

Pinney, Chris C., *The Illustrated Veterinary Guide*, TAB Books, 1992

Pinniger, R.S. (ed), *Jones' Animal Nursing*, Pergamon Press (1972)

Plummer, D. B., *Ferrets*, The Boydell Press (1992)

– *Tales of a Rat Hunting Man*, The Boydell Press (1978)

Primack, Richard B., *Essentials of Conservation Biology*, Sinauer Associates Inc. (1993)

Quesenberry, Katherine E. & Carpenter, James W., *Ferrets, Rabbits and Rodents: Clinical Medicine and Surgery*, W.B. Saunders Company (2003)

Schreiber, A., Wirth, R., Riffel, M. and Van Rompaey, H., *Weasels, Badgers, Civets and Mongooses and their Relatives*, IUCN (International Union for Conservation of Nature and Natural resources) (1991)

Simpson, Gillian, *Practical Veterinary Nursing* (3rd edn), British Small Animal Veterinary Association (1991)

Sleeman, Paddy, *Stoats & Weasels, Polecats & Martens*, Whittet Books (1989)

Southern, H.N., *The Handbook of British Mammals*, The Mammal Society (1979)

Taylor, Fred J., *The Shooting Times Guide to Ferreting*, Buchan and Enright (1977)

— *Guide to Ferreting*, Buchan and Enright, (1988)

Wellstead, Graham, *The Ferret and Ferreting Guide*, David and Charles (1981)

Young, J.Z., *The Life of Vertebrates*, Oxford University Press (1978)

Appendix 4

Useful Addresses

The following were all correct at the time of going to press; however, details and officers do change, especially where the organisation is a voluntary one. In all cases, where a reply is required, always enclose a stamped self-addressed envelope with all postal enquiries.

The inclusion of an organisation does not imply any standard or the author's approval of that organisation; neither does the exclusion of any organisation imply any disapproval.

Denmark
dansk ilder forening
www.ilder.dk

Finland
suomen frettiyhdistys ry
www.lilaslair.com

Germany
Frettchen Club Berlin
c/o Christiane Friese
10725 Berlin
030-4617193

Frettchen und Marderclub Deutschland
c/o Stephanie Larisch
30539 Hannover
0511-9507023

Frettchen Zuchtverband
Dorf Strasse 26
0-1401 Teschendorf
www.frettchenzuchtbund.org
Frettchenfreunde NRW
In der Grossen Heide 6
44339 Dortmund
www.frettchenfreunde.info

Netherlands
Stichting de Fret
Boteronstraat 41
1445 LH Purmerend
Holland
www.fret.nl

Norway
Norsh Ilder Forening
www.ilder.no

Sweden
Svenska Illervänners
www.illervanner.se

Stif Vast/Svenska Tam-Iller Foreningen
Norra Barsjov 19
427 35 Lindome
www.svenskatamillerforeningen.se

This organisation has many local groups, details of which can be obtained from the above address.

United Kingdom
British Association for Shooting and Conservation
Marford Mill
Rossett
Wrexham
Clwyd LL12 0HL
www.basc.org.uk
Tel 01244 573 000

An organisation which, although not specifically for ferrets and ferreters, does work to encourage and maintain shooting sports; as such, ferreting interests are represented within the Association.

Countryside Alliance
The Old Town Hall
367 Kennington Road
London
SE11 4PT
www.countryside-alliance.org.uk
Tel: 020 7840 9200

An organisation which represents the interests of all country sports, including ferreting.

Deben Group Industries
Deben Way
Melton
Woodbridge
Suffolk IP12 1RB
www.deben.com
Tel: 01394 387762

Manufacturers of ferret detectors and other ferreting equipment.

The Ferret Roadshow
Holestone Gate Road
Holestone Moor
Ashover
Derbyshire S45 0JS
Tel 01246 591590
www.ferret-school.co.uk
info@ferret-school.co.uk

Formed by the author in 1982, the Ferret Roadshow travels the UK giving displays and lectures on all aspects of ferrets and ferreting.

The Gem Ferret Care Group
www.gemferret.co.uk

The National Database of Ferret Friendly Vets
The National Ferret School
Holestone Gate Road
Holestone Moor
Ashover
Derbyshire S45 0JS
Tel 01246 591590
www.ferret-school.co.uk
info@ferret-school.co.uk

A computer database of veterinary surgeons, established by the author, who have experience or interest in ferrets and their ailments. Addition to the database is free.

The National Ferret School
Holestone Gate Road
Holestone Moor
Ashover
Derbyshire S45 0JS
Tel 01246 591590
www.ferret-school.co.uk
info@ferret-school.co.uk

This organisation, established in 1988, runs courses on all aspects of ferrets and ferreting. Fact and information sheets are published by, and available from, the school, as well as two Codes of Practice – one for ferreting operations, and the other for organisers of ferret racing and displays (see Appendices 5 and 6).

The School's Ferreting Code Of Practice remains the only such code to have been officially approved by the UK's leading country sports organisations – the BASC, the Countryside Alliance and the Game & Wildlife Conservation Trust.

UK Rabbit Management Services
Holestone Gate Road
Holestone Moor
Ashover
Derbyshire S45 0JS
Tel 01246 591590
www.uk-rabbit-management.co.uk
info@uk-rabbit-management.co.uk

Established in 1981, UKRMS is the UK's leading rabbit management service, which specialises in eco-friendly and traditional techniques. The company also runs courses on all aspects of rabbit management.

Universities Federation for Animal Welfare
8 Hamilton Close
South Mimms
Potters Bar
Hertfordshire EN6 3QD

There are many ferret clubs in the UK, but as the contact details vary from time to time, it has been impossible to give details in this book. Those wishing to find their local ferret club should make a search of the internet using a good search engine.

United States
The American Ferret Association
PO Box 3986
Frederick
MD 21705
www.ferret.org

The American Ferret Veterinary Association
1014 Williamson Street
Madison
WI

The Black-footed Ferret Fund
c/o Wyoming Game and Fish Department
5400 Bishop Boulevard
Cheyenne
WY 82002

For information on the progress of the black-footed ferret breeding and re-introduction project, and also to make donations to help the work of protecting the domestic ferret's most endangered relative.

The Ferret Fancier's Club
713 Chautauqua Court
Pittsburgh
PA 15214

Ferret Unity and Registration Organisation (FURO)
PO Box 844
Elon College
North Carolina
27244

Ferret World Inc
6 Water Street
Box 555
Assonet
Massachusetts 02702

Manufacturers and distributors of a wide range of ferret equipment.

The International Ferret Association
PO Box 522
Roanoke
Virginia
24003

Marshall Pet Products
5740 Limekiln Road
Wolcott
New York
14590

The largest breeders of ferrets in the USA.

The Miami Ferret Club Inc.
19225 SW 93rd Road
Miami
Florida
33157

Appendix 5

The Ferreting Code of Practice

INTRODUCTION

Properly conducted, ferreting – the use of ferrets to bolt rabbits from their warrens – is both a sport and one of the most effective methods of controlling the wild rabbit population. It causes little disturbance to the environment and results in healthy fresh meat for the table, and income for local people. Ferreting was the way in which many of us were introduced to the delights of country sports, and its importance to farmers, landowners and sportsmen and women should not be underestimated.

However, as with all pastimes – and particularly those involving country sports – all practitioners have to consider public reaction to their actions and the public face of their practices. It is important, therefore, that we all conduct ourselves in a truly acceptable manner at all times. It is hoped that this Code of Practice will help enshrine our finest traditions and ensure the highest standards from all who participate in this ancient and fascinating sport. This Code was produced in conjunction and coopera-

Six rabbits will eat as much as one sheep, making them an agricultural pest which needs controlling.

tion with the BASC, the Countryside Alliance and the Game Conservancy, and is the only such Code to have achieved approval by all of these leading country sports organisations.

With the increase in the UK's rabbit population, farmers and landowners are fighting a constant battle against this pest. No one wishes to see the elimination of the rabbit, but the damage that it does to agriculture is colossal. Six rabbits eat as much food as a sheep, and there are currently an estimated 60 million-plus rabbits in the UK. Most people would agree that it is more sensible, humane and 'environmentally friendly' to control this burgeoning rabbit population by means of ferreting than by using poisons, gassing or by the introduction of diseases specifically intended to destroy the rabbit population *en masse*.

When the UK rabbit population reached epidemic proportions (over 105 million) in the early 1950s, myxomatosis was introduced by humans, and the disease killed 95% of the rabbit population of England and Wales within six months. While there can be no argument that this reduced the rabbit population, the pain, distress and suffering that it caused has shocked many a true country person.

Even within the relatively small area of these islands, there are many variations – including climate and crops – which will affect ferreting, but it is the spirit of this Code that is important. All true sportsmen and women should encourage its use, support those who adhere to it, and draw it to the attention of all interested parties.

THE RULES

1.   All ferreting operations must be conducted in strict accord with the laws relating to all aspects of the sport, and it is the duty of all ferreters to understand and observe these laws.

      The laws which apply to ferreting are many and varied, including those protecting property, the person, and the welfare of animals. All ferreters should understand the reasons for these laws, the importance of observing them, and the consequences of breaking them. The old maxim that ignorance is no excuse should be borne in mind at all times.

      Ferreting is also a sport and, even where the intention is to control the rabbit population, the best traditions of country sports should be upheld and practised.

2.   Ferreting must only be carried out with the permission of the landowner or duly authorised tenant over whose land you intend to ferret. Operations should be limited to the areas, times and dates stipulated by that person.

      It is a wise precaution to have permission in writing, and this should be carried at all times while out ferreting over that area. It is also common sense to inform the landowner/tenant when you will be working a specific area, and check that this is convenient. Be prepared to avoid any areas if requested, and stick to the letter of your agreement. This is particularly important during the game shooting or hunting seasons.

3.   As far as possible, ferreting should be limited to times of the year when there are fewest litters of rabbits, normally from September to the middle of

March (although this may vary with geographical location of site).

It is in no one's interests to ferret burrows where many of the does will have litters.

4.　　At least one member of the ferreting team should have considerable experience of the sport, and all involved should be competent in the humane killing of captured rabbits.

Ferreting can be a complicated exercise, and considerable skill, born of experience, is needed to carry out the work efficiently and effectively. Practitioners should acquire these skills and experience under the watchful eye of seasoned ferreters.

All members of the ferreting team should be able to identify signs of badgers, which are highly protected and the deliberate disturbance of which is illegal. No ferreting or disturbance of any kind should be carried out in or near badger setts, even if there are signs of rabbits in the same workings; neither should any ferrets be entered into any workings were it is believed that foxes are living, as this is potentially dangerous to the ferrets.

Any dead rabbits found in suspicious circumstances should be reported to the landowner or tenant, as should any signs of poaching. The National Ferret School issues an information sheet on rabbit calicivirus (also known as rabbit viral haemorrhagic disease – RVHD) in wild rabbits.

5.　　Whenever possible, ferreting should not be carried out single-handedly. When digging or working in excavations, there is always a risk – however slight – of a lone ferreter becoming trapped or buried alive. The risks are minimised when ferreters work in teams, while this also leads to more efficient and effective rabbit management

Even where small teams go ferreting, it is advisable to inform a responsible person of the exact location(s) of operations, and give an expected time of return. Where ferreters are unavoidably delayed, every effort should be made to inform this person, in order to avoid the unnecessary involvement of the emergency services.

6.　　A basic first aid kit should be carried by every ferreting team, and at least one member of the team should have experience and/or training in elementary first aid procedures.

The first aid kit, which will be suitable for use on both human and ferret injuries, and can easily be carried in the net bag, should contain the following:

Assortment of bandages
Cotton wool or gauze
Sticking plasters
Antiseptic liquid, ointment or powder
Tweezers
Scissors

The basic principles of first aid are to save life, prevent the condition worsening, and aid the recovery of the victim, and are applicable to both humans and animals. The order in which injuries should be addressed is summed up by the ABC Rule:

A – An open AIRWAY
B – Adequate BREATHING
C – Sufficient CIRCULATION

Many organisations run courses in human first aid techniques (e.g. St John Ambulance, St Andrew's Ambulance Association, and the British Red Cross). It is highly recommended that ferreters attend a first aid course; this could save the life of a member of the team.

7. Veterinary care should be sought at the earliest possible time for any injured ferret, dog or hawk used in ferreting.
    Even where first aid has been administered correctly, some injuries will still require veterinary attention. Large wounds will need sutures, while other wounds and bites may require the administration of suitable antibiotics or other prescription-only medicines.

8. Dogs and hawks used with ferrets must be well trained and accepting of ferrets.
    Dogs and hawks, both of which can be and often are used in conjunction with ferrets, have the potential to inflict grave injuries on ferrets, and so they must be extremely well trained and have experience of ferrets, in order to minimise the risks of injury to all parties. When a ferret is attacked, it can defend itself by inflicting painful wounds on the attacker – be it dog or hawk – and even the hand of a human intent on helping the beleaguered ferret.
    No dog or hawk should be released (slipped) on to quarry until the rabbit is completely clear of the burrow entrances; this will reduce the risk of ferrets being injured, and also of injured rabbits escaping down burrows.
    The safety and well-being of dogs and hawks is as important as that of the ferrets, and should be given a high priority.

9. Where the intention is to shoot bolting rabbits, Guns must not shoot until the rabbit is in clear view, and well clear of the burrow opening. It is all too easy for ferrets to be accidentally shot by those who ignore this simple, common-sense rule.
    Many ferrets give hard chase to their quarry and are often very close behind the rabbit. It is bad practice – and extremely dangerous – to shoot without clearly identifying one's target. Injured rabbits, shot close to the entrance to their burrow, may kick themselves down the burrow, rendering it impossible for ferreters to dispatch them humanely.
    In the UK, little owls (*Athene noctua*) often live in burrows, and it is not

unusual for one to fly away from a rabbit burrow during ferreting operations. It is obvious that this protected raptor should not be shot at.

Guns should be placed in such a way as to avoid any risks of accidents to other Guns, while at the same time giving them every opportunity to shoot bolting rabbits.

10.   Remove all nets from the burrow after the completion of ferreting operations.

The most commonly used nets for ferreting are purse-nets, and these are often made of material which is difficult to see when the net is in position over the opening to a rabbit burrow. Ferreters should count the number of nets placed (not an easy task, especially when several operatives are working at the same time), and ensure that the same number are collected in. By using a rubber band to secure each wound-up net stored in one's net bag, and ensuring that these are put on the nets after they have been used, it is easy to see if there are any nets not yet picked up, i.e. there will be one or more 'spare' rubber bands. It is good practice to check the whole area before leaving or moving on to another burrow.

Nets used for ferreting should be checked and serviced regularly, with any damage being repaired before use.

11.   Any damage to land or property caused by your ferreting activities should be repaired before you leave the area.

It is possible for fences to be damaged slightly, or holes (sometimes substantial) to be dug during a day's ferreting, and it is only honourable and fair to repair all damaged items to their original state, if at all possible. Holes not only look unsightly, but are a potential danger to wildlife, livestock and humans. Unless the landowner or tenant specifically requests otherwise, all ferreted holes should be filled in (back-filled). This action serves three worth-

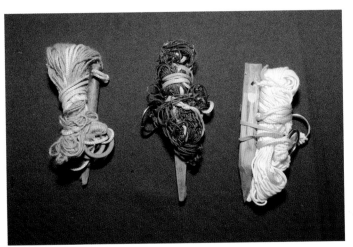

When carried in the net bag, all nets should be wrapped and stored neatly, to prevent tangling and facilitate easy extraction of a single net from the bag when needed. Note rubber bands round each net to secure them.

while purposes; it shows clearly which holes have been ferreted, dissuades rabbits from returning, and shows areas in which rabbits have returned.

All ferreters should possess adequate third party liability insurance; we recommend a minimum of £1 million. It is also recommended that ferreters should have suitable personal accident insurance. These types of insurance are offered to all members of the British Association for Shooting & Conservation (BASC) and the Countryside Alliance, as part of their membership package.

12.   No ferret should be worked muzzled or coped.

It is unnecessary and potentially dangerous for a ferret to be worked with either a muzzle or a cope in place; this applies to devices made from string, leather, metal, plastic or any other material. It is only a matter of time before working ferrets meet a rat in the burrow and, if the ferret is fitted with a muzzle, it will have no way of defending itself. A cornered rat will attack the defenceless ferret, inflicting serious injury or even death.

The barbaric practice of breaking or removing a ferret's teeth is illegal, immoral, and entirely needless. Provided that a ferret is not starved, it is highly unlikely, in the event of an underground kill, to eat the dead rabbit and stay underground to sleep off the meal. Ferrets used for ferreting do not have to be vicious or nasty; on the contrary, everyone will fare better by using 'quiet' (tame and manageable) ferrets at all times.

13.   Wherever possible, every ferret used in rabbiting operations should be equipped with an electronic ferret finder. Where this is not possible, the ferreters should have with them a trained, traditional liner ferret, and at least one member of the ferreting team should have considerable experience in the use of such an animal.

The easy availability and efficiency of the modern electronic ferret finder should be utilised to find ferrets which have killed underground, become trapped or for other reasons fail to return to the surface. In order to ensure that this equipment operates properly, it should be serviced regularly, and have new batteries fitted at regular intervals and always after extensive usage. The equipment should be checked before leaving base on a ferreting foray; any equipment not working 100% should not be used, but sent to a qualified person for repair.

Where it is not possible to equip every ferret with a transponder, an 'electronic liner' should be used. This is the modern equivalent of the old-fashioned liner but, instead of an actual line, the ferret is equipped with a transponder. Apart from this modernisation, its use is exactly the same as a traditional liner, i.e. to bolt out ferrets that have laid-up, and then remain with the dead rabbit until it is dug out by the ferreter.

Should electronic detectors break down or cease to function correctly, it is useful for ferreters to have the knowledge and skills to use a traditional liner, and we recommend that all ferreters learn these skills in advance of such a situation arising.

14.   Ferrets which do not leave the burrows or otherwise become 'lost' must not be left.

It is likely that some ferrets, entered into a rabbit burrow during rabbit management operations, will not come out of their own accord before it is time for the human operatives to leave the area. Every effort must be made to find these animals. The use of an electronic ferret detector will help in this task, and ferreters must be willing to dig to extricate trapped or lost ferrets. Obviously, suitable grafts, chads or spades should be carried by the ferreting team.

An escaped ferret can cause havoc on a formal game shoot, devastating stocks of pheasants and other game birds, while farmers who keep chickens or waterfowl will soon realise why the French gave the ferret its name of *poulet chat* (chicken cat). Under the Animals Welfare Act 2006, it is illegal to abandon an animal in this way.

Where extensive efforts to find the ferret in the burrow have been unsuccessful, other efforts must be made to recapture the ferrets, and live-catch mink traps are ideal for this purpose. One or two should be taken on every ferreting trip, although they may be left in the vehicle until, and unless, they are needed. Traps should be baited with pieces of food, and placed in close proximity to the area in which the missing ferret was last seen or heard. These traps must be inspected every few hours before dark, at dawn the next day, and at regular intervals until the ferret is found or hope abandoned; the latter should not happen until every possible effort has been made to recapture the ferret.

Landowners and tenants in and around the area of the loss should be notified of any missing ferrets, and given a 24-hour contact number to use should the ferret be seen or captured. Where a third party contacts the ferreters to inform them of the capture of a lost ferret, the ferreters should retrieve the ferret within 24 hours. A small reward should also be offered along with sincere gratitude; this also has a public relations angle, and is well worth the small investment for the goodwill that such a gift engenders.

15.   No fewer than two ferrets should be taken on any rabbiting trip, and no ferrets should be overworked.

Ferrets, like all animals, will tire during working, and are then tempted to find a comfortable place in which to rest; this could easily lead to lost ferrets, and much work on the part of the ferreters. It is far better to use ferrets in teams, and alternate between them, allowing each regular rest periods. Any ferret which shows obvious signs of fatigue should not be used.

16.   All ferrets used for rabbiting should be fed before work and if necessary during the day.

It is unnecessary and unwise to work hungry ferrets, since they may be too weak to keep going all day and, being hungry, are always tempted to kill and eat any rabbits that they find. All ferrets used for rabbiting should be fed as normal until the day of the operation. A light meal first thing on the morning of the ferreting trip will ensure that they have enough energy to work, will not be so full that they are lethargic, and are not tempted to kill and eat any of the rabbits they are hunting. When you take a break for sustenance, allow the ferrets a light snack. Do not forget to allow ferrets to have access to clean drinking water at regular intervals throughout the day.

17.   Ferrets should be transported in carrying boxes, and not bags, throughout rabbiting operations.

Strong wooden carrying boxes, with adequate ventilation, are safer and far more comfortable for the ferrets. Bags offer little protection against rain, cold, heat or a size 9 boot. The boxes should be filled with dry straw; the amounts should be varied according to prevailing weather conditions and temperatures. Carrying boxes containing ferrets should not be left in the sun, and care must be exercised when placing them in the car for transport to or from the ferreting venue. No ferret should be left inside a locked car for more than a few minutes.

All boxes should be equipped with a strong, safe method of securing the opening or lid, and preventing the escape of the ferrets, and this should be used whenever a ferret is in the box.

18.   When ferreting on railway embankments or the verges of main roads, all ferreters should wear highly visible clothing.

For obvious safety reasons, it is vital that drivers can see anyone who may have strayed onto or near roads and railway tracks; the high visibility jacket is the ideal way of ensuring high visibility. On minor roads, a simple mesh waistcoat may suffice, but near busy main roads, dual carriageways, motorways and railway tracks, it is recommended that high visibility jackets are worn by all ferreters working in such areas. You may feel rather self-conscious wearing such a jacket, but it is common sense to protect yourself in this way.

19.   Throughout a day's ferreting, the ferret's welfare must be foremost.

At the end of the day, all ferrets should be fed a top-quality meal of reasonable proportions, dried (if necessary) and checked for injuries. Relevant first aid treatment should be administered if necessary, and then the ferrets should be placed in a safe and secure cage or box, which should then be placed in a safe position in your vehicle, before attending to yourself. If necessary, injured ferrets should be taken to a veterinary surgeon.

Once back at base, the ferrets should be removed from their carrying box, checked again, and when certain of their well-being, placed back in their own cub or court, along with a fresh meal.

Dogs and hawks used in conjunction with the ferrets should receive similar treatment and care, again before you tend to your own needs.

20.   Ferreting relies on the goodwill of landowners and tenants, as well as the efficient, effective and thoughtful operations of ferreters, and this goodwill must be nurtured.

Ferreters should treat all land and property with respect, and honour all agreements made between themselves and the landowner or tenant over whose land they operate. Guests should not be taken onto ferreting land without the express prior permission of the landowner or tenant. Carried out correctly and with courtesy, ferreting should result in a symbiotic relationship, where ferreter, landowner and tenant benefit from a good working relationship. Good ferreters will be recommended to other landowners and tenants.

SUMMARY

This Code of Practice attempts to give outlines on common-sense ways of ensuring the well-being of all involved in ferreting – both animals and humans. Responsible ferreters will already be adhering to the principles set out in this document, but it is hoped that, as well as guiding newcomers to the sport, 'old hands' will find this Code a reminder of the importance of conducting oneself in an appropriate manner at all times.

Newcomers are also advised to read books on the subject, and to receive adequate training; the National Ferret School runs courses on ferrets, ferreting and related subjects (including the use of dogs, hawks and ferrets together); the School also produces fact sheets, books, information and DVDs on all aspects of ferrets and ferreting.

Further information can be obtained from the National Ferret School, Honeybank, Holestone Gate Road, Holestone Moor, Ashover, Derbyshire S45 0JS; telephone 01246 591590.

E-mail – info@ferret-school.co.uk
If requesting information via the postal service, please enclose a large (A4) sae, with suitable stamp attached.

Appendix 6

Code of Practice for Organisers of Ferret Racing and Displays

Ferret displays, involving racing and other activities, are now extremely common – and rightly popular – at all major country shows and other events throughout the British Isles, and in many other countries. It is important to remember that these ferret displays are the public image of ferrets and ferret keeping, and it makes sense, therefore, to ensure that ferret keepers, clubs and others with an interest in the subject are seen in the best light.

All too often, ferrets used in displays are not treated correctly and, as a result, both ferrets and members of the public are put at risk – sometimes even injured. All it takes is one serious incident, and all ferret owners will suffer the consequences. This Code of Practice consists of common-sense practices and precautions which are designed to ensure that accidents do not occur – to either ferrets or humans – and that ferrets used in displays are treated correctly. It is in everyone's interests that all adhere to this code.

The Organisers of Ferret Racing and Displays should:

1.   Ensure that the well-being of the ferrets is paramount and is given the highest priority at all times; no ferret should be used too often, nor made to endure any treatment that may result in injury or stress, to either ferret or human.

2.   Ensure that the ferrets are transported to and from the event in suitable caging, with due regard to temperature regulation, and in full compliance with all Transport of Animals Orders. An Animal Transportation Authorisation is required by law.

3.   Ensure that, except when actually participating in racing or similar activities, the ferrets have constant access to suitable food, water and shelter.

4.   Ensure that, except when actually participating in racing or similar activities, the ferrets are kept in cages of appropriate size and construction, allowing shelter from extremes of temperature and prevailing weather conditions.

5.   Ensure that the ferrets receive regular veterinary treatment and vaccinations, as necessary.

6.   Ensure that the ferrets are accompanied at all times while on site, and one person should be given overall responsibility for the ferrets' welfare while on site; their guidance should be sought before any activity involving the ferrets.

7.   Be registered under the Performing Animals (Regulation) Act 1925.

8.   Hold public liability insurance to a minimum of £1M.

9.   Ensure conformance with health and safety guidelines, and adherence with the Code of Conduct of the Council for Environmental Education.

10.   Ensure that at least one member of the team on site holds a recognised first aid qualification, and that suitable first aid equipment (for both animals and humans) is available at all times.

For further information, please contact the National Ferret School, Honeybank, Holestone Gate Road, Holestone Moor, Ashover, Derbyshire S45 0JS. Tel. 01246 591590.
www.ferret-racing.co.uk   info@ferret-racing.co.uk

Index